NUMERICAL WEATHER PREDICTION

Venkata Bhaskar Rao Dodla
IMS Fellow
Professor of Meteorology (Retd.)
Andhra University

CRC Press
Taylor & Francis Group
Boca Raton London New York

CRC Press is an imprint of the
Taylor & Francis Group, an **informa** business

INDIAN METEOROLOGICAL SOCIETY
Mausam Bhavan Complex, Lodi Road, New Delhi - 110 003.

BSP **BS Publications**
A unit of **BSP Books Pvt. Ltd.**
4-4-309/316, Giriraj Lane,
Sultan Bazar, Hyderabad - 500 095

First published 2023
by CRC Press
4 Park Square, Milton Park, Abingdon, Oxon, OX14 4RN

and by CRC Press
6000 Broken Sound Parkway NW, Suite 300, Boca Raton, FL 33487-2742

CRC Press is an imprint of Informa UK Limited

British Library Cataloguing-in-Publication Data
A catalogue record for this book is available from the British Library

ISBN: 9781032406251 (hbk)
ISBN: 9781032406268 (pbk)
ISBN: 9781003354017 (ebk)

DOI: 10.4324/9781003354017

Typeset in Times New Roman and Bookman Old Style
by BSP Books, Hyderabad 500 095

BSP BOOKS

Contents

Foreword

Indian Meteorological Society is a non profit scientific society established in 1957 for advancement of Meteorological and allied sciences and dissemination of the knowledge of such sciences both among the scientific workers and among the public and their application for societal benefits. The society organizes National and International Symposiums (TROPMET /INTROMET), Regional Seminars and Workshops and has constituted a number of awards to promote research in meteorology. The IMS popularizes meteorology by organizing public lectures, teachers and media workshops interactive sessions with students, essay, debate and quiz competitions on popular meteorological topics, During interactions it was felt that books for general public written in easy to understand format on various advance topics of meteorology are not available in India. To meet this demand, it is decided by IMS to bring out popular book series as a part of its Diamond Jubilee Celebrations.

Weather predictions have shown steady improvement in recent decades leading to change in public perception about weather forecasts. Today weather forecasts are helping in saving lives by early warning of extreme weather like cyclone, heat wave etc and improving productivity by weather advisories to agriculture, energy, health, aviation, marine and other weather sensitive sectors. Common people are using weather forecasts in planning their day to day activities. Numerical Weather Prediction (NWP) techniques using high power computing and observations from Weather Satellites and Radars have played key role in improving skill of forecasts. NWP Techniques are being used in all spatial and temporal scales of forecasts starting from Nowcasts to seasonal forecasts and climate projections. Somehow, knowledge about NWP has been confined to NWP Centers and some academic and research institutes. Books on NWP essentially cater to research community. Need to have a popular book on Numerical Weather Prediction was being felt for quite some time. Therefore, IMS decided to berin out second book in its popular book series on Numerical Weather Prediction.

The book under IMS Popular Book series on 'Numerical Weather Prediction' is authored by Prof D.V. Bhaskar Rao pioneer in Numerical Weather Prediction Meteorology in the country. He initiated teaching and research programs in NWP at Andhra University in 1977 and made significant research contributions on tropical cyclone modelling. With his strong passion for teaching and research, he encouraged and mentored numerous students of M.Sc., M.Tech., and Ph.D. programs inducing them to a career in meteorology and allied disciplines. I am thankful to Prof Bhaskar Rao for having agreed to IMS request to write the book. The book is written in simple, easy to understand format and compressively covers all aspects of Numerical Weather Prediction. The book covers Numerical Methods, Hierarchy of Atmospheric Models, Objective Analysis, Parameterisation, Ensemble Prediction and current status of NWP Models. I am also thankful to Prof. Someshwar Das of Rajasthan Central University for reviewing the manuscript.

I am sure readers will find the book interesting and will be better informed about NWP. The book will also create interest in **Numerical Weather Prediction** among the students.

AVM (Dr) Ajit Tyagi
President
Indian Meteorological Society

Preface

Weather influences on earth's humanity finds a reference in ancient Indian scriptures, such as description of hydrological cycle in "Bhagavad Gita" (~3000 B.C.) and of the seasonal cycles, cloud and rain processes in "Brihatsamhita" by Varahamihira (~500 A.D.). A verse from Bhagavad Gita (Chapter 3, Verse 14) reads as

annad bhavanti bhūtāni
parjanyād anna sambhavaḥ
yajhad bhavati parjanyoyajhaḥ
karma samudbhavaḥ

The gist of this verse is "Living beings are created and find their source in food, food is created by rainfall, Parjanya (Rain God) gives us the material blessings of rainfall and thus abundance in the material world by the creation of food and other crops".

Weather plays a very important role in the lives of all humanity, due to its impacts. Mankind is vulnerable to weather extremes such as extremes of temperature (cold and heat waves), extremes of rainfall (drought and floods), extremes of wind (storms such as cyclones, thunderstorms, tornadoes etc.). So awareness to weather changes and need to know of their causes had prompted human instinct to strive towards understanding the atmospheric changes and learn to protect from weather perils. Understanding weather as continuous changes in the atmosphere providing sometime comfort and sometimes distress seems to have been a part of learning in different parts of the globe since ancient times as evidenced by the writings in the ancient scriptures. The quote above taken from "Bhagavad Gita" is supposed to be given in 3137 BC depict the seasons, cloud, rain and hydrological cycle which reflect the scientific thought of those times.

Current scientific thought accepts that the earth's atmosphere system is driven by energy from Sun, in the form of radiation. In brief, the scientific processes that contribute for the continuously changing atmosphere are as follows. Solar radiation heats up the earth-atmosphere system differentially, corresponding to rotation of earth around its own axis once in 24 hours and around Sun once in 365.25 days causing day and night and seasons. The radiation energy passes through the atmosphere and most part of it reaches the earth surface and the surface of the earth gets heated depending on the type of surface (basically land, water or ice and land featuring innumerable types) leading to different temperatures at different places. These horizontally varying temperatures give rise to pressure variations and the horizontal pressure differences lead to movement of air from one location to another (generally called as wind). The atmosphere, striving to reach an equilibrium state as per nature, is in continuous motion in all the three dimensions of space. Though looks simple, the changes in the atmosphere result from complex interactions between the land, water and ice surfaces of the earth with atmosphere above. An attempt to delineate these changes brings out the simultaneous presence of atmospheric motion at different time and spatial scales, (i.e.) from the smallest scales of molecular motion, turbulent motion to synoptic and large scale variations where the smallest spatial scales have smallest time scales and the time scales tend to be longer as the spatial scales tend to be larger. The enormous complexity of the presence of innumerable variations makes the understanding of the atmospheric system very difficult. But the characteristic behaviour of the various systems could be understood through observations and theory as they have inherent periodic nature. It is important to realise that it would be impossible to predict a random behaviour and the basis of the weather prediction is the deterministic characteristic of the atmosphere. The basic premise for studying weather should be towards continuously improving our understanding.

Weather prediction had been the focus of attention ever since the science of meteorology has been carved out of the science of physics. The first chapter of this monograph describes the development of weather prediction since 1900s. Despite the complexity of the weather prediction problem, profound advancements have been during the last few decades mainly due to the rapid advancements in the observation systems and computational resources to support weather prediction using atmospheric models. The development of the atmospheric models based on fundamental physical laws and solved using well understood numerical methods have established numerical weather prediction as the state-of-art methodology. Technological advancements in Radar and satellite measurements and high performance computing have helped the orientation of weather prediction from subjective and qualitative style to objective and quantified mode. The weather predictions are currently oriented to suit different users from agriculture, transport, industries, and sports etc. apart from the general public who plan their daily chores following the weather predictions. In recent times of looming world energy crisis, when strong efforts are being made towards harnessing alternate energy such as wind and solar energy, weather prediction plays an important role to provide quantified data at the required time and spatial scales to serve the particular demand.

This monograph has been prepared to provide a basic understanding of the numerical weather prediction to a general reader with knowledge of basic physics and mathematics. The stream of thought follows a lucid style with least use of technical nomenclature. Since the entire subject of numerical weather prediction is based on mathematics and physics, elaboration of mathematics could not be avoided while describing the formulation of models (chapter 2) and numerical methods (chapter 3). Information on supplementary topics that aid these two topics is given as appendices. Any reader who is not keen about the mathematics may skip the mathematical part of these two topics and follow the general information on the other topics. As such most of the relevant information has been segregated and presented as about modeling, numerical methods, objective analysis and data assimilation methods to generate initial conditions and parameterization of physical processes. The chaos theory of Lorenz, which originated from his experiments on weather prediction, has been presented for the benefit of understanding the deterministic predictions and uncertainties of numerical weather predictions. Above all, this monograph should help a student and researcher to understand the intricacies of not only the weather prediction but also the current problems and strategies. The purpose of this monograph deems to be fulfilled if the information fills the gap between the available books and research articles and help find clarifications as needed.

-Author

Acknowledgements

The author thanks the Indian Meteorological Society, and it's Current National President Dr. Ajit Tyagi, for the opportunity to prepare this monograph. The content of this monograph is mainly drawn from the author's lecture notes of the last four decades and teaching of numerical weather prediction at Andhra University has definitely helped. The author thanks his present faculty colleague, Dr. G. Ch. Satyanarayana for his help in the preparation of the manuscript into the required format and shape.

-Author

1. Advent of Numerical Weather Prediction

The dawn of numerical weather prediction corresponds with the proposition of the "laws of motion" by Newton (1642-1727), of which the relevant law states that "The rate of change of momentum of a body is equal to the sum of the forces acting on the body". This law is the basis of weather prediction. The scientific achievements in fluid dynamics have been applied to atmosphere such as of Euler's (1707-1783) equations for fluid motion in incompressible, non-viscous flows to represent the changes at specified locations in space and of Claude-Louis Navier (1785-1836) and George Gabriel Stokes (1819-1903) who described the motion in viscous fluid with the addition of a diffusive viscous force term. Cleveland Abbe (1838-1916) was the first to recognise the application of hydrodynamics and thermodynamics to the atmosphere, and to propose a mathematical approach to weather forecasting.

Vilhelm Bjerknes (1862–1951)

Vilhelm Bjerknes (1862–1951), Norwegian scientist, was the first to postulate the possibilities of weather prediction and identify the need for detailed observations in 3-dimensional space and application of appropriate physical laws. Bjerknes recognised pressure, temperature, density, humidity and three components of velocity as the seven basic variables and the seven independent equations as the three equations of motion, the continuity equation, the equation of state and the equations for the two laws of thermodynamics. Bjerknes, being aware of the unattainable analytical integration, perceived a procedure based on graphical charts representing the initial state and estimating their future evolution.

Exner (1908) came up with a different methodology to calculate the pressure changes resulting from advective process, using mean zonal wind derived from observed temperatures and with the constrained assumptions of geostrophic balance and constant thermal forcing. His calculations of pressure changes were realistic and were recognised as a first attempt at systematic weather forecasting.

Lewis Fry Richardson (1881-1953) was the first to attempt numerical weather forecasting through a direct solution of the equations of motion using finite difference method. Richardson, who had been working on graphical methods and the development of finite difference method for a different application in 1910, considered its application of these methods for weather forecasting in 1911. But he got interested in weather prediction problem only in 1913, when he joined the Meteorological Office in the Southern Uplands of Scotland. He learnt of the weather forecasting problem enunciated by Bjerknes and realising the possibilities of applying finite difference methods for solving the differential equations came up with a precise computational scheme for the Bjerknes prognostic component. Richardson was aware of the complexities and enormity of the numerical computation and impracticality of its immediate application but could foresee its real time weather predictability with future developments in algorithms and computational speed. His dream had come true in 1954 with the first real time NWP experiment at Princeton, USA. Although Richardson started his work on weather prediction in 1913 and initiated his computations to generate pressure changes, the progress was constrained due to his work as an ambulance driver for about 2 years in France during the world war period of 1916-1918.

Richardson used the seven differential equations formulated for atmospheric processes to represent the state of the atmosphere through seven variables: pressure, temperature, density, water content and velocity components eastward, northward and upward as of Bjerknes. Richardson divided the 3-dimensional atmosphere into 60000 cube elements with a resolution of 3^0 in east-west, 200 km in north-south and 5 cells in the vertical covering the globe. The variables were defined at the centre of each cell, the differential equations were approximated in finite difference form, and the rates of change of the variables were then calculated. Richardson calculated the pressure changes in 6-hours over central Europe. Perhaps Richardson might have spent about 1000 hours in about 2 years to produce this forecast. Unfortunately his forecast was registered as a catastrophic failure as his calculations yielded a pressure change of 145 hPa in 6-hours.

Richardson (1922) published a book "Weather Prediction by Numerical Process", in which he narrated the details of his research. This book was acclaimed by many, and Sir Napier Shaw regarded it as 'a magnum opus on weather prediction'. Richardson identified the reasons for the failure of his calculations as related to large, spurious convergence values that resulted from errors in the wind observations, large grid distance, and the interpolation process.

F. J. W. Whipple, perhaps was the first to understand and suggest the reasons for Richardson's unrealistic forecast as due to the presence of gravity and sound waves, which move faster than meteorological phenomena as solutions of the numerical equations that were used. This argument was later realised to be true and not the reasons related to sparsity of observations. Richardson's supposition of the large distances between observations as the cause of spurious convergence estimates is not true as the error results from the compensating terms in the convergence equation.

The calculations of Richardson show large disparity in the magnitudes of the pressure gradient and Coriolis force terms, which should have been of equal magnitude and opposite sign. Sir Napier Shaw was fully aware of the atmosphere tending to restore a balance of pressure gradient and wind velocity and the transition takes place in infinitesimal stages implying the changes in the balance also to be infinitesimal which is now known to be geostrophic adjustment. Richardson provided more emphasis on controlling the spurious divergence and neglected the role of ageostrophic wind component.

Max Margules (1904) cxamined the predictability of pressure changes using the conservation of mass principle and identified that synoptic-scale pressure changes cannot be estimated due to the possibilities of errors srising from compensating terms in the equation of continuity.

Rossby (1898-1957) had made profound influencing contributions to both applied and theoretical meteorology and oceanography. His theoretical results have shown that the large scale and synoptic scale changes in the atmospheric circulation could be predicted by considering readjustments of the horizontal velocity field neglecting the changes in the vertical structure of the atmosphere. His theoretical results have segregated the high-frequency gravity-inertia waves with phase speeds of up to hundreds of metres per second and large divergence and the low frequency motions with phase speeds of the order of ten metres per second and characteristic periods of a few days which are of meteorological significance. Rossby's theoretical work on atmospheric waves had been the basis of the development of atmospheric models for weather prediction. The geostrophic theory enunciates the dominance of rotational mode containing the bulk of energy over divergence in large scale atmospheric flow and the weak interaction between the fast speed and slow moving waves. Rossby and his collaborators carried out the first numerical prediction experiment in 1940s using a "single-layer barotropic atmosphere" which had set the initiative for the development of forecasting theory and techniques through comparisons of calculated and observed atmospheric states.

During 1940s, four significant advancements had helped the formulation of models for weather prediction. These are (i) the pioneering works of Jack Bjerknes, Sutcliffe on the dynamics of development of middle latitude systems; Rossby's description of atmospheric waves; Kuo's work on barotropic instability and Charney's work on baroclinic instability which improved the understanding of the dynamics of pressure systems and waves, (ii) radiosonde observations that improved the initial state, (iii) Courant's work on the stability properties of finite difference schemes and the inference of the Courant, Friedrichs and Lewy stability criterion, and (iv) the development of automatic electronic computing machines that could enhance the computational speed.

The contributions of Charney towards NWP were phenomenal. Charney postulated the physical basis for numerical prediction of large-scale atmospheric motions, and the filtering of the high frequency waves that would allow numerical prediction up to 2-days. This filtering of meteorologically unimportant sound and gravity waves helped with the choice of larger time step satisfying the CFL stability criterion and had been the basis of a hierarchy of filtered models that were used for NWP in 1950s and 1960s.

Charney (1947) postulated the importance of quasi-geostrophic balance which indicates the forcing mechanisms and dominant modes of hydrodynamic instability and of the manner in which energy is dispersed and dissipated in the atmosphere. Observations show that the bulk of the energy in the troposphere is in rotational modes with advective time scales, and the forcing of gravity waves is weak due to the longer time scale of the external solar forcing than the inertial time scale, implying that the coupling between the rotational and gravity wave components is also weak. This quasi-geostrophic theory has been the basis for success of numerical weather prediction in 1950s.

Charney (1948) published his analysis on the scale of atmospheric motions which lead to the mathematical elimination of gravity waves through geostrophic assumption which lead to postulation of the quasi-geostrophic system. Charney (1949) advocated the use of barotropic vorticity equation with the assumption of constant static stability following Rossby (1939) as the simplest of the hierarchy of the filtered quasi-geostrophic models that would help further development of baroclinic models as needed for the prediction of developing pressure systems (such as cyclones).

Independently, Arnt Eliassen (1949) also developed semi-geostrophic equations with the use of geostrophic assumption in the acceleration equations and was the first to transform the equations of motion to isobaric coordinate system, which had become important in subsequent applications in the numerical modelling of the atmosphere.

Charney, Fjørtoft and von Neumann (1950) made their first numerical forecast using the non-divergent barotropic vorticity equation. This forecast was obtained using the first ever electronic digital computer ENIAC built in 1945 at Maryland, USA. It was a huge machine with 18,000 thermionic valves, massive banks of switches and large plug boards with tangled skeins of connecting wires, filling a large room and consuming some 140 kW (kilo Watt) of power. Program commands were specified by setting the positions of a multitude of 10-pole rotary switches on large arrays called function tables, and input and output was by means of punch-cards. Charney et al used the barotropic vorticity equation transformed to polar stereographic projection, discretised to a grid of 19 × 16 points which corresponded to 8 degrees longitude at 45oN (736 km at the North Pole and 494 km at 20oN), adopted centered spatial finite difference and leapfrog time differencing schemes.

Electronic Digital computer ENIAC

This was the first successful experiment which led to the development and use of different quasi-geostrophic models for weather forecasting in middle latitudes. Since the filtering concept is based on the hydrostatic and geostrophic approximations, the forecasts were constrained as the approximations may not hold good all over the globe, especially that the geostrophic approximation holds good where Coriolis force term had enough magnitude equal to the pressure gradient force.

Prof. Charney
(1917-1981)

Model initial conditions were at 500 hPa level and the model integrations were performed for 24-hours keeping the values at the lateral boundaries constant.

Fjørtoft, returned to Norway in 1951 after being a part of the first successful NWP and not having a computer to continue his work on NWP, developed a graphical method (1952) to integrate the barotropic vorticity equation. The computations of vorticity advection and the solution of Helmholtz equation were performed graphically completing a 24-hour forecast within 3-hours. A repetition of the Princeton experiment using ENIAC has shown that the graphical method produced results comparable with those using the ENIAC. Fjørtoft's method had been used by several Air Force weather forecast centres in 1952 and 1953. Although this manual method was in use for a brief period, it is historically important as this was the first effort to use Lagrangian advection method and also as the first successful effort to establish the graphical methods proposed by Vilhelm Bjerknes.

The first computer based 72-hour predictions using real time observations suitable for operational use were made at the International Meteorological Institute at Stockholm University, Sweden in November, 1954 using a Swedish computer "BESK" (Persson, 2005). As the results of trials were encouraging, regular operations have been started in 1956. The Swedish work of operational objective forecasting using barotropic vorticity equation was useful for almost a decade.

The occurrence of a severe storm in November, 1950 over the east coast of the United States, and the incapability of the barotropic model to predict it have brought out the need for development and application of models with barioclinic dynamics. Subsequently several baroclinic models were developed all based on quasi-geostrophic system. The numerical prediction group at Princeton reported good simulation of the November-1950 severe storm using 2- and 3-level models. Thus, the operational use of baroclinic models in operational forecasting had started.

The successes of weather prediction with the baroclinic models lead to the formation of a Joint Numerical Weather Prediction Unit in USA in July, 1954 combining the units of Air Weather

Service, the Naval Weather Service and the Weather Bureau. The unit had George Cressman as Director, Joseph Smagorinsky as head of the operational section, Philip Thompson as head of the development section and Art Bedient as head of the computer section which had started operations in 1955. Here it had transpired that some of the earlier successes with the quasi-geostrophic baroclinic model was only a chance and that the multi-level model consistently fared worse than the barotropic version (Shuman, 1989). Due to this reason, operational forecasting reverted to the use of single-level model from 1958.

Phillips (1990) had presented an interesting analysis that the numerical prediction of Charney et al in 1954 was successful because of the choice of a large geographical area although the grid resolution had to be coarse to satisfy the CFL stability criterion. He commented that had the experiment was confined to a smaller area with finer resolution, it would have been a failure attributing the scientific reason to the spread of errors across the lateral boundaries. In this context, use of multi-level models meant smaller forecast domain that will have increased forecast errors propagating through the lateral boundaries due to the retention of the initial values at the boundaries.

At those contemporary times, the limitations of using multi-level quasi-geostrophic models were noted in Britain and Germany. Phillips (2000) contented that the use of "stream function" in the place of "geopotential" could have improved the predictions, which may have been true to some extent but the limitations on the applicability of geostrophic approximation in the tropics returned the entire procedure back to the use of the original governing equations of the atmosphere as employed by Richardson.

Research with the primitive equations began at NMC (now NCEP) in 1959. Charney, realising the problems with quasi-geostrophic models, initiated use of a linear barotropic primitive equation model containing the CFL stability criterion. Initial work of Charney had shown that the predicted motion field consisted of both the Rossby and gravity wave motion which had helped to explore the feasibility of integrating the primitive equations. The use of a six-level primitive equation model was initiated in June, 1966, running on a CDC-6600 (Shuman and

Hovermale, 1968) which had shown considerable improvement in the forecast skill. Platzman (1967) compared the NMC primitive equation model with that of Richardson and reported more similarities than differences although the NMC model had been independently developed using Eliassen's (1949) formulation of equations in isobaric coordinates.

Karl-Heinz Hinkelmann of Germany had advocated the use of primitive equations citing that the atmospheric dynamics and energetics would be simulated more realistically than the filtered equations. He made the first systematic attempt to generate suitable initial conditions through control of high frequency oscillations by taking up appropriate initialization and argued that the larger computational time due to smaller time step would be offset by simpler algorithms free of the solution of elliptic equations in the quasi-geostrophic models. Hinkelmann's first prediction experiments with the primitive equations had yielded a very good simulation of the evolution of extra-tropical frontal system. This success was the prelude to the initiation of operational numerical forecasting in the Deutscher Wetterdienst in 1966 (Reiser 2000), and this was the first ever use of the primitive equations in an operational environment.

At the beginning, the primitive equation models (Smagorinsky, 1958; Hinkelmann, 1959) were adiabatic, with dry physics. Later an equation for moisture has been introduced into the primitive equation system and its introduction had caused numerical problems with the rapid development of small scale convective systems arising from grid scale conditional instability. To overcome this problem, the concept of "moist convective adjustment" had been imposed that would supress gravitational instabilities. Subsequently this had been the starting point for developing parameterization theories for cumulus scale convection.

In Britain, the concept of using single-level models for weather forecasting was not accepted. Only in 1972, a ten-level primitive equation model (Bushby and Timpson, 1967) incorporating parameterisation of physical processes of heat, moisture and momentum through the bottom boundary, topographic forcing, sub-grid scale convection and lateral diffusion had been started which yielded credible forecasts of precipitation. Although

Britain was initially hesitant, the UK Met Office held a leading position in the development of numerical weather prediction.

Initialization: At the time when the application of atmospheric models had been shifted to the use of primitive equation models, more attention was drawn towards the problem of initialization. When primitive equations are used for numerical prediction, any imbalance between the mass and velocity fields in the initial state would reflect as anomalously large gravity-inertia waves that may persist for a long time yielding to spurious forecasts. It was the presence of such imbalance in the initial fields that gave rise to the totally unrealistic pressure tendency of 145 hPa/6h obtained in Richardson experiment.

Balance in the initial data is achieved by a process noted as initialization, in which the initial fields are presented in such a way that the amplitude of the gravity inertia waves remains small throughout the forecast. If the fields are not initialized the spurious oscillations will occur leading to forecast errors.

The atmospheric balance is subtle, in the sense that minor perturbations may disrupt it but robust that local imbalances tend to be removed through dispersion by gravity-inertia waves leading towards natural adjustment between the mass and wind fields.

Numerical forecast experiments with and without initialization have shown contrasting differences with the absence of spurious oscillations and realistic tendencies in the initialised prediction process. The initialization was achieved through static and dynamic strategies.

Hinkelmann (1959) derived the initial winds using the geostrophic relation, and integrated the primitive equations using a very short time step. Although Hinkelmann's experimental forecasts with the primitive equations and initialization were noted to be far superior to the quasi-geostrophic model integrations, Charney (1955) proposed that the use of nonlinear balance equation may produce a better estimate of the initial wind field, as this equation includes the curvature of the streamlines implying the wind as non-divergent. The difficulty with the solution of the balance equation using the geopotential and stream function fields was the constraint that the data satisfy an ellipticity relation. To overcome this problem, Phillips (1960) suggested that an improvement of balance would result if the vertical velocity field is derived using quasi-geostrophic ω–equation and the equation of continuity. Each of these methods had rendered improvements but the noise problem persisted.

Sasaki (1958) formulated an initialization method based on variational calculus. In this method, both the wind and mass fields were constrained to fit a balance condition while remaining close to the original analysis by a suitable choice of weighting functions, with more weightage for the wind in low latitudes (tropics) and more weightage for the height in high extra-tropical latitudes. Although this variational method of initialization was not widely used, this method is now the basis of modern data assimilation techniques. All the above methods are the static initialization methods.

In the approach of dynamic initialization, a forecast model itself was used to define the initial fields (Miyakoda and Moyer, 1968; Nitta and Hovermale, 1969). The supposition is that the dissipative processes damp out high frequency noise as the forecast evolves. In this process, numerical integration schemes having selective damping of high frequency components can be utilised. In one of the approaches, the model is integrated forward one time-step and then backward to the initial time with the dissipation action completing one cycle and the cycle is repeated enough times till the high frequency components were damped out. The forecast starting from this initialization is noise-free, but had the problem of damping the meteorologically significant motions as well as the gravity waves.

In another approach, the initial fields are segregated into normal mode components, the gravity-inertia waves are filtered retaining the slow moving rotational waves (Dickinson and Williamson, 1972). Although this process of 'linear normal mode initialization' would assure a noise-free forecast, it was noted that the noise reappeared due to nonlinear interaction of the slow waves leading to appearance of gravity waves. So the problem of noise essentially remained. Machenhauer (1977) presented the method of "'nonlinear normal mode initialization" to control the growth of gravity waves, in which the initial rate-of-change of the gravity waves were set to zero and this method remarkably helped, as the model forecasts initialised this way were very smooth and without the appearance of the spurious gravity wave oscillations. Baer (1977) developed a similar method based on more rigorous mathematical reasoning but it was more difficult to use and so the Machenhauer's method was the most popularly used.

However, there were problems to use the normal mode initialization method in the case of limited area models; wherein the normal modes at the lateral boundaries cannot be derived. Daley (1991) provided detailed discussion on the various methods of initialization such as the bounded derivative method, the implicit normal mode method and the Laplace transform method.

A new approach, the method of digital filter initialization (DFI) was introduced by Lynch and Huang (1992). This method uses an optimal filter. The method is similar to that applied in signal processing, in which low-pass, high-pass and bandpass filters are generated and applied. Essentially for the initialization problem in weather prediction, a filter is applied to preserve the low-frequency oscillations from contamination by the high-frequency oscillations. This method is now being used for application in limited area models such as WRF etc.

The development of NWP in different countries was mainly dependent on the availability of computational resources and the knowledge of the atmospheric model developments. For example, the weather changes are linked to large-scale quasi-geostrophic regimes in the middle latitudes, sub-synoptic scale features related to topography and convection dominate the weather in the tropical regions. Thus, the simple models such as barotropic models and quasi-geostrophic models which could be run on the computer systems of 1960s had helped development of NWP in middle latitude regions, where the use of primitive equation models suitable for prediction over tropics requiring computer systems of 1980s was needed and hindered the application of NWP over tropical regions.

While the above methods are related to dynamic initialization, Krishnamurti (1991) proposed a novel method of physical initialization, in which the vertical atmospheric structure is adjusted to the observed rainfall rates through the convection parameterization scheme and the method was found to significantly improve numerical weather prediction in tropics. While there is no doubt that the development of NWP was confined to USA in the first two decades of 1960s and 1970s, other countries soon started application of NWP and contributed to the development of their regions.

Japan: Japanese scientists have made many important contributions to the development of NWP. Prof. Shigekata Syono, a Professor of Meteorology at Tokyo University during 1945-1969, was not only responsible for the progress of NWP but also for mentoring many researchers who have made invaluable contributions in Japan and USA. Syono and his group published their first research on one-dimensional barotropic forecasting in 1950 in the *Journal of the Meteorological Society of Japan* which indicated similarity of their ideas with those of Charney's group in USA. Kanzaburo Gambo, one of Syono's students, worked at the Institute for Advanced Study in Princeton (IAS) with Charney's group for two years during 1952-1954 and got updated with the progress of NWP in USA. Between 1954 and 1960, the Japanese meteorologists performed their numerical calculations using Fjørtoft's graphical methods, desk calculators and a small relay-switching computer (FACOM 100) to produce predictions of precipitation, tropical cyclone movement.

The Japan Meteorological Agency (JMA) started their NWP activities in March 1959, with the acquisition of IBM-704 computer and soon after started giving 48-hour predictions using a hemispheric barotropic model operationally.

The organisation of an international symposium on NWP at Tokyo in 1969 (7-11 November) helped the scientists of USA and Japan with the exchange of ideas and this was the meeting where the need for development of primitive equation models had been recognised and Ed Lorenz presented his results on the divergence of nonlinear computations due to round-off errors.

Some of the students of Syono, due to limited opportunities in Japan at that time, migrated to USA in 1960s and made significant contributions to atmospheric modeling and NWP. To mention a few of them are: Akio Arakawa, Syukuru Manabe,

Kiku Miyakoda, Michio Yanai, Yoshi Sasaki, Akira Kasahara, Yoshio Kurihara whose contributions are now well known to meteorologists all over the world.

Germany: Like Japan, Germany also lacked the resources to be at the forefront of European meteorology. The Deutscher Wetterdienst (DWD) (German meteorological service) was in the US-Zone after World War II, where Hermann Flohn was the head of the research division. Karl-Heinz Hinkelmann, of the meteorological service, visited the Swedish Meteorological Institute in 1951 where he worked on barotropic model calculations under the guidance of C G Rossby. Hinkelmann started working on baroclinic models soon after his return. In 1952 DWD was created by West Germany and Hinkelman had started model integrations using graphical methods. By 1954, access to computers had helped produce barotropic model forecasts up to 72-hours using first the Swedish computer BESK, and later an IBM 704 in Paris. Although Hinkelmann viewed the application of hydrostatic primitive equation models for NWP, the same could not be realised till a CDC3400 computer was acquired in November 1965. At DWD, operational barotropic forecasts for 72-hours started in October 1966, and baroclinic forecasts started in 1967 with the arrival of a computer CD 3800. In the former German Democratic Republic, the acquisition of a Soviet computer BESM-6 helped initiate routine NWP in January 1971. Later the GDR meteorological service ran a 4-level with 300 km resolution, to generate numerical forecasts up to 72 hours.

France: France was one of the first few countries whose national meteorological organisation evinced interest in NWP as early as 1952. In 1954, Guy Dady was sent to Stockholm for a few months to work with Rossby's group. After his return, Dady was permitted a few hours access to one of the three computers (available in France at that time), which he used to generate predictions up to 72 hours with a simple barotropic model at a grid resolution of 400 km. Although the Met´eorologie´ Nationale (now Meteo France) acquired its own computer KL901 in November 1960, the progress of NWP in France was slow due to the political turmoil till 1964. Daniel Rousseau, a junior scientist to Dady, was sent to MIT, USA in 1965 where he completed his Master's program and thesis on general circulation model under Prof. Charney's guidance. On 1

December 1967 the Met´eorologie´ Nationale acquired a new computer, CDC 6400, and operational NWP with a single filtered barotropic model was started in 1968. Subsequently in 1970, experiments with balanced and multi-level primitive equation models had been initiated along with operational predictions with a filtered model. Robert Sadourny and Olivier Talagrand, had received training at UCLA in 1965 and started working on data assimilation and global models. The purchase of the CDC6400 in 1970 helped the French scientists to develop primitive equation models. The NWP division, led by Guy Dady and Rousseau developed a 10-level primitive equation model with 36-km grid resolution suitable for small-scale studies, and since 1972 a research version with 180-km grid resolution was run on daily basis to supplement the operational NWP with a filtered model.

Algeria: The Algerian Meteorological Service, with the support from WMO, had acquired a small computer and started NWP activity. Dr. Coiffier with help of Dr. Jean Lepas, France had started a numerical model set up which included Cressman objective analysis and a filtered model.

Belgium: The group of Jacques van Miegham, J. van Isacker and Defriese started work on a barotropic model in 1955 using a computer at Antwerp. Operational NWP forecasts for Western Europe have started in 1962 using a barotropic model at 500-km grid resolution covering the hemisphere. Later a 2-parameter model was used. Belgium was one of the countries that supported the establishment of the ECMWF.

Italy: The Italian Meteorological Service started NWP in 1955, and the first daily forecasts were produced in 1959-60. Sabino Palmieri, who received training in NWP for two years at the Joint Numerical Weather Prediction Unit, Maryland, USA, was the first to introduce and develop a barotropic model in 1960, for the Italian Meteorological Service with the available limited computer time of about 1 hour/day. The model integration area covered half of the hemisphere and required 1-hour computer time for 2-day integration. A Numerical Forecast Centre was established in 1967, and as the Italian Meteorological Service acquired an IBM 360/30 in 1968, a two-level quasi-geostrophic model replaced the barotropic model.

Later the model was upgraded to 4-level quasi-geostrophic model in 1970 as the computer was upgraded to IBM360/40.

Canada: Canada had contributed to the development of atmospheric models. NWP was first initiated in the mid-1950s at the University of Toronto. Dr. Michael Kwizak, of the Canadian Meteorological Service, was the one who initiated NWP in Canada. Andre´ J. Robert joined Dr. Kwizak in 1959 and the two of them developed NWP. Their efforts resulted in starting operational NWP using a non-divergent barotropic model on 1 September 1963. This group had developed a four-level baroclinic model on the 28 × 32 grid, and later enhanced the same to a 51 × 55 grid which was used for NWP. Dr. Robert was responsible for the development of global spectral model in Canada which had been the basis of present day global spectral models.

Australia: Australian meteorologists started NWP as early as 1950s. The meteorology groups at the Melbourne University initiated NWP education and later advancements followed at the Bureau of Meteorology. Dr. Ross Maine of the Melbourne University reported their work on Charney- Eliassen one-dimensional model as early as 1957. Teaching and research in NWP was started at the Department of Meteorology in 1958, and Mr. Jensen made a Master thesis "Numerical forecasting with the barotropic model", in which 24-hour predictions on a 21 x 17 grid with 300-km resolution were reported. Later Jensen and Radok produced their first barotropic predictions in 1960 using the UTECOM computer facility at the University of New South Wales.

The Bureau of Meteorology took up NWP actively since 1963, started their operational forecasts in 1969 using a barotropic model with a 24x36 grid with a resolution of 254 km, and hemispheric forecasts in 1973.

New Zealand: The NWP experiments covering the Southern Hemisphere had been carried out in New Zealand first using the Fjørtoft graphical method, and later using a barotropic model run on an IBM 650. The predictions were of stream function using quasi-geostrophic barotropic model on a 12x18 longitude-latitude grid.

Israel: NWP was started in 1962, giving 72-hour forecasts of 500 hPa using a geostrophic barotropic model run on a Philco 2000.

USSR: USSR started NWP following the Bergen school of thought and reported to have come up with numerical model calculations by Kibel and Blinova in 1960s. The development of NWP was not well followed due to political situation in erstwhile USSR and also that, much of the scientific material was in Russian language. Reports indicate that solution of quasi-geostrophic model using graphical methods was started in 1954, and the operational forecasts using a barotropic model started in 1959. The Hydrometeorological Service of the USSR, in 1962, reported the use of a 3-level quasi-geostrophic model with 26x22 grid points and a resolution of 300 km for 48-hour forecasts, and a primitive equation model on experimental basis twice a week. In Leningrad a 2-level quasi-geostrophic model, and in Novosibirsk a 5-level quasi-geostrophic model were used for NWP.

Czechoslovakia: A small group of scientists under the leadership of S. Brandejs started their research in NWP at the Meteorological Institute of Charles University in Prague as early as 1952. By 1963, the Meteorological Institute of Charles University (headed by S. Brandejs), the research department of the Central Hydrometeorological Institute (headed J. J´ılek), and the department of atmospheric circulation at the Meteorological Laboratory of the Czechoslovak Academy of Sciences (headed by V. V´ıtek) had initiated NWP research work at their laboratories. Initial NWP research on barotropic models was carried out using a Soviet computer, Ural 1. Routine operational forecasts with a barotropic model (domain of 24x20 and 315-km resolution) started in 1966 using a British-made computer LEO 360. In 1966 Michal Bat'ka worked with Professor Brandejs on a model using the omega equation. The scientists struggled through the times of cold war and Soviet occupation periods to move forward in NWP finally establishing them to have a consolidated NWP department in Central Hydrometeorological Institute where the ALADIN model is being run now on a small NEC-SX6 machine.

People's Republic of China: According to the information available, NWP had started in 1954 with the use of a 2-layer model being solved using graphical methods. Barotropic model forecasts for 48-hours period using a computer were initiated in 1960. During early 1960s, research on quasi-gestrophic baroclinic models and later on primitive equation models had been reported.

Finland: NWP had its roots in the Department of Meteorology, University of Helsinki. Dr L. A. Vuorela, was the first to take up NWP in the early 1950s as a visiting scientist in Sweden and the USA. The first NWP effort was put up by Daniel Soderman and Juhani Rinne using a barotropic model (18x20 grid points) on an IBM 1620. Operational NWP forecasts with a 3-level filtered model were started in 1970 at the Finnish Meteorological Institute under the leadership of Dr Eero Holopainen, acting Chief of the Weather Forecasting Section.

Denmark: The progress of NWP in Denmark was in three phases, first in mid-1950s, second during 1960-70 and the third in mid-1980s. During 1950s, Ragnar Fjørtoft at the University of Copenhagen developed the graphical method for solving the barotropic vorticity equation, which was used till the computers became available. Aksel Wiin-Nielsen, Hans S. Buch and Harry van Loon, who worked with Fjørtoft on this project joined other institutions and contributed to NWP. Ole Lang Rasmussen, at the Danish Meteorological Institute, started NWP producing 48-hour forecasts (32x40 grid points at 300-km resolution) with a barotropic model. In 1973, forecasts with a baroclinic model (25x23 grid points, 254 km resolution) were started. The Nordic countries (Denmark, Finland, Iceland, Norway and Sweden) had initiated a coordinated project on NWP in 1984, under the leadership of Bennert Machenhauer, to develop a High Resolution Limited Area Model (HIRLAM). The HIRLAM had largely contributed to the progress of NWP in the NORDIC countries.

Netherlands: NWP started in mid 1960s at the Royal Meteorological Institute of the Netherlands (KNMI) under the leadership of F. H. Schmidt. Operational numerical weather forecasts were issued using a 3-level baroclinic model at a resolution of 375 km.

Norway: The Norwegian Meteorological Institute (DNMI) started their operational NWP forecasts in 1962, releasing two sets of

48-hour predictions, one using a barotropic model applied at the 500 hPa level and the other by a baroclinic model to derive 300-700 hPa thickness. Norway did not join the ECMWF initially but joined later with minimal participation. Three Norwegians, Einar Hoiland Ragnar Fjørtoft and Arnt Eliassen were famous for their significant research contributions but they could not promote progress of NWP in Norway for different reasons. Hoiland dedicated to theoretical hydrodynamics, Fjørtoft restricted to his graphical methodology, and Eliassen despite his association with the first NWP experiment at Princeton confined himself to research (famous contributions of Eliassen–Palm flux and Sawyer–Eliassen theories) at his university leaving NWP to the meteorological organisation.

Spain: Spain evinced interest in NWP around 1960 but could not progress due to paucity of computational resources. In 1965, a NWP group was formed as a part of the National Meteorological Service (NMS) whose collaboration with the Swedish Meteorological and Hydrological Institute helped their progress with NWP. The meteorological service acquired its first computer, IBM 360/340 in 1966-67 and started producing NWP forecasts using a hemispheric barotropic model. The National Meteorological Institute (formerly NMS) developed the Swedish limited area model based on the ECMWF global model in 1984 due to the upgradation with a Fujitsu computer.

Yugoslavia: Operational NWP with a barotropic model was started in 1970 using IBM 360/44, generating 96-hour predictions of 500 hPa geopotential (330 km resolution) covering the areas of Europe, Atlantic and North America. Two research groups at the Federal Meteorological Institute contributed to NWP, one used a two-parameter model and the other a primitive equation model. Mesinger and Janjic had developed the first successful primitive equation model in 1976, which later became popular as the Hydrometeorological Institute and Belgrade University HIBU) model.

Ireland: The Irish Meteorological Service first used the Swedish quasi-geostrophic model running it on an outside computer, but with the acquisition of a DEC20/50 computer in 1979 switched to the Yugoslav five-level primitive equation model.

Indian Scenario: Meteorology in India had started with the setting up of the first observatory at Madras (now Chennai) in 1802 by the East India Company. Meteorological observations were started at different locations gradually over the next decades. The establishment of the India Meteorological Department (IMD) in 1875 was the dawn of meteorology in India. The work at IMD was mostly confined to the establishment of weather observatories till about 1945 (i.e.) the end of World War II, as all efforts were devoted to aviation services. Numerical weather prediction in India was initiated by Dr. P.K. Das. The first ever numerical forecasting experiment was carried by Dr. Das in 1958, wherein he attempted to predict a monsoon depression using a geostrophic barotropic model and solving the Laplacian equation by relaxation method manually as electronic computers were not available at that time in India.

IMD could not keep pace with the scientific developments in atmospheric dynamics and scientific computing outside due to dedication of all IMD services to aviation sector during the two world war periods. Atmospheric research in universities and academic institutions was yet to start in India. Prof. P. K. Das, who had joined the IMD in early 1950s, took the initiative and leadership of NWP in India. The establishment of "Institute for Tropical Meteorology (presently the Indian Institute of Tropical Meteorology) in November 1962, exclusively to promote meteorological research, had helped initiation of research relevant to Indian region which included numerical weather prediction. Prof. P.R. Pisharoty, the founding Director of ITM and also a student of the renowned American Professor J. Bjerkenes, acquired the first electronic computer IBM 1620 for meteorological research. This had helped many of the Indian meteorologists to step into NWP.

Dr. Prasad Kumar Das (1926-2005)

The first research efforts at ITM were on the barotropic and baroclinic instabilities of Indian summer monsoon using quasi-geostrophic models by Dr. S. Daggupaty and Dr. VBRao. Dr. G. C. Asnani, a student of Prof. Pisharoty, started working with quasi-geostrophic barotropic models for weather prediction in 1967. Three of the ITM scientists visited the National Meteorological Centre (NMC) in USA during 1967, who received training in objective analysis, satellite meteorology and NWP. During this period, they had the opportunity to attend special lectures in NWP by Dr. Takahashi Nitta of Japan, who was then a visiting scientist at the NMC, which generated their interest in the NWP. As a part of the visiting program, Sikka was introduced to satellite data inputs to numerical analysis and prediction (SINAP). Dr. Shukla visited Japan Meteo-rological Agency, Tokyo for two months on his return from US where he worked with a two-level primitive equation model under the guidance of Prof. K. Gambo and Dr. T. Nitta, and the results of his study were later presented at the NWP symposium in Tokyo in 1968.

Due to encouragement given to young scientists by Dr. Saha, the then Head of the Forecasting Division since 1967, significant research efforts were made by Asnani on the barotropical model with geopotential as input; Mr. Shukla on the non-divergent barotropic model with stream function as the predicted variable; Mr. Ramanathan on the objective analysis methods and Mr. Dixit on the derivation of geopotential field from winds using the reverse balance equation.

As work on NWP at ITM progressed, the institute's IBM 1620 computer was found inadequate for NWP, ITM scientists were permitted to use the CDC 3600 computer at Tata Institute for Fundamental Research (TIFR) for their NWP work. In 1970, Mr. Shukla and Mr. Sikka reported the first successful integration of the non-divergent barotropic model with winds, the results of which were appreciated by Prof. P. Koteswaram, the then Director General of IMD. Subsequently Dr. Ramanathan started work on the primitive equation (PE) barotropic model and Mr. Sikka on a five-layer quasi-geostrophic model with geopotential derived from winds. By 1974, Dr. Ramanathan was successful in using limited area primitive equation barotropic model for NWP over India, and Mr. H. S. Bedi, a meteorologist at IMD, successfully integrated a global barotropic model. After Shukla's departure for USA in 1971, Mr. S. S. Singh took his place and started working on a multi-level PE model. Mr. Singh was sent to Japan for training in NWP models, and he published results of his joint study with Dr. M. Sugi on the prediction of monsoon depressions in 1980. During the period 1966-1986, several researchers (a few of them: Mr.

S. Rajamani, Dr. S. K. Mishra, Dr. G. C. Asnani, Dr. R. N. Keshvamurthy, Mr. S. T. Awade) made diagnostic and prognostic studies using quasi-geostrophic models and balanced models. Most importantly, all the computer codes were prepared by the concerned researchers themselves which helped the learning and application of numerical modeling for weather prediction.

India Meteorological Department: Dr. P. K. Das's affiliation with the IMD in 1950 marks the beginning of NWP era in India. Ever since, he had contributed enormously for the growth of NWP not only at IMD but also at IITM, IIT-Delhi, Andhra University and various national organisations and academic institutes. Dr. Das, with education and training in meteorology at Imperial College, London and due to his keen interest in atmospheric dynamics, initiated NWP at IMD. His earliest contribution was in 1957 on the numerical prediction of a monsoon depression using a barotropic model. He had to use graphical methods for the solution of the model equations as computers were not available. There was not much progress for a decade due to lack of computer facility. The IMD group headed by Dr. Das started using the available computers in New Delhi at that time, an IBM 1620 computer at the Planning Commission, the ICL 1901 computer at Delhi University and a computer at IIT-Delhi. Using these facilities, the small NWP group of R.K.Datta, B.M. Chabra and B.V.Singh successfully produced numerical predictions with a quasi-geostrophic barotropic model and a multi-level quasi-geostrophic model and made them operational in 1970 and 1975 respectively. IMD acquired their first computer, IBM360/44 in 1973 and the NWP progressed rapidly thereafter. The IMD group streamlined NWP by incorporating suitable methodologies for data acquisition, quality control, objective analysis and model integration. The shift of Dr. Y.Ramanathan and Dr. H.S. Bedi from IITM to IMD and the joining of Dr. M.C. Sinha by 1975 consolidated the NWP group. Simultaneously, improvements on the operational multi-level quasi-geostrophic model, development of a multi-level primitive equation model and generating model diagnostics had consolidated the NWP applications. IMD had provided excellent operational weather prediction support, using multi-level quasi-geostrophic limited area model, during MONEX (Monsoon Experiment) in 1979. Dr. P.K.Das had designed a model and simulated the storm surge associated with a severe tropical cyclone that struck Bangladesh in 1971 which had been the beginning of the modeling of cyclone-induced-storm surges at IIT-Delhi in India.

The NWP efforts at IMD were stimulated due to the Indo-US Science & Technology Initiative (STI) launched in 1985, under which Dr. Bohra visited Florida State University (FSU) to learn about the regional primitive equation model and Dr. Basu had been exposed to NMC's R40L18 global spectral model at the National Meteorological Centre (now NCEP) in USA. Later the FSU regional model was implemented at IMD in 1987. Under the same program, Dr. M.B.Mathur and Mr. Bansal visited NMC in 1987 to work on quasi-Lagrangian model for tropical cyclone prediction and optimum interpolation method for objective analysis respectively.

The European Centre for Medium Range Weather Forecasting (ECMWF), established in 1974 for the development and application of medium-range weather forecasting models in Europe, facilitated participation of researchers from several countries to collaborate in their development efforts. As a part of this, Dr. R. K. Datta worked on "the prediction of monsoon flow over the Indian sub-continent by the ECMWF model" during a visit to the ECMWF in 1985, the results of which were published as a Technical Memorandum (No.113) of ECMWF in 1986. Subsequently Dr. A. Hollingsworth of ECMWF visited IMD in 1986 and collaborated with Dr. Datta and his group at IMD to evaluate the ECMWF model's performance for predicting characteristic weather situations over India. This has provided the IMD scientists an exposure to the global atmospheric models for weather prediction applications.

The India Meteorological Department had made rapid strides in the development of weather predictions on different time and spatial scales to meet national requirements. A High Performance Computing System (HPCS) with peak speed of 14.2 Tera Flops was commissioned in January 2010 to run different models. The HPCS receives all data from manual and automatic devices across the globe, processes it and generates global and regional forecasts. The Global Data Assimilation (GDAS) cycle is used for assimilation of all available observations and run 4 times a day (00, 06, 12 and 18 UTC). IMD had adopted the Global Forecast System (NCEP, USA) at

T382L64 resolution and now runs it in real-time mode for generating 7-day predictions. IMD currently runs several models operationally, the Limited Area Model (LAM), MM5, WRF and Quasi-Lagrangian Model (QLM) models for short range prediction. The MM5 model is run at the horizontal resolution of 45 km; WRF model is run at 9 km resolution both models using boundary conditions from NCMRWF T-254 Global Forecast System (GFS). The QLM model is specifically used for cyclone track prediction over North Indian Ocean. IMD is also making nowcasting, on experimental basis, using "Warning Decision Support System Integrated Information (WDSS-II)", software developed by National Severe Storms Laboratory, USA. With the ingesting of Indian DWR observations, the application software is capable of detecting storm cells and removing anomalous propagation echoes. IMD is currently producing weather parameters for generating agro-meteorological advisories at district level, using model outputs from five different global models – (i) NCMRWF- T254, (ii) ECMWF-T799, (iii) JMA-T859, (iv) UKMO and (v) NCEP GFS T-382 considering them as ensemble members. IMD in cooperation with IITM is predicting the monsoon weather on extended time scales of 7 to 30 days using four variants of CFSv2 coupled model, which are (i) CFSv2 at T382 (~38 km) (ii) CFSv2 at T126 (~100 km) (iii) GFSbc (bias corrected SST from CFSv2) at T382 and (iv) GFSbc at T126. The Multi-model ensemble (MME) out of the above 4 suite of models are run operationally for 32 days once a week (with Wednesday initial conditions) with 4 ensemble members (one control and 3 perturbed) each for CFSv2T382, CFSv2T126, GFSbcT382 and GFSbcT126 and average ensemble forecast anomalies that are computed using the 4 sets of model runs of 4 members will constitute the ensemble prediction.

National Centre for Medium Range Weather Forecasting: In January 1988, Government of India approved the establishment of National Centre for Medium Range Weather Forecasting (NCMRWF) on similar lines of ECMWF due to its success in Europe, with the main objective of providing medium-range (~7 days) weather predictions for preparing agriculture advisories to 127 agro-meteorological field units of India. With this mission, NCMRWF has grown to be a "Centre of Excellence", continuously striving to improve NWP systems through research, development and applications.

Government of India acquired the first super computer, Cray XMP-14 (One processor and four million words memory) vector machine for use by NCMRWF for weather prediction in India, which was formally inaugurated by Mr. Rajiv Gandhi, the then Honorable Prime Minister on March 25, 1989.

Due to the mandate of providing medium range weather predictions, need for use of global atmospheric models had become imminent. Initial rapid development of NCMRWF operations were due to the strong support from Centre for Ocean Land and Atmosphere Interactions (COLA) and National Centers for Environmental Prediction (NCEP). Due to the support from Prof. Shukla and his COLA team, their R40 model had been adopted for experimental forecasting at NCMRWF during 1990-1994. Prof. Kanamitsu, NCEP helped with the installation and use of their T80 model which became operational at NCMRWF in 1993. NCEP's support has been continuous providing access to their advanced analysis assimilation system, ocean wave forecast system, mesoscale ETA Model System and in 2006, NCEP provided their T254L64 model which is used for operational NWP since June 2007.

Support also came from the Australian Bureau of Meteorology, the Japan Meteorological Agency (JMA) and the U.K. Met Office for the medium range forecasting efforts at NCMRWF. Thus, effective support from various organisations and individual scientists helped NCMRWF to rapidly advance from the application of simple limited area models to sophisticated global models at the operational level. Concurrently NCMRWF achieved success in the other important components of starting Agricultural Meteorology Field Units (AMFU), streamlining communications, downscaling the variable information to the desired AMFU level, and preparing advisories involving agricultural scientists.

Due to the importance for data in NWP, procedures have been developed for data quality control and data assimilation towards improvement of initial conditions. At NCMRWF, during the initial periods, it was realised that the mass and momentum fields were not well balanced in the tropics which prompted the development of assimilation strategies such as the use of empirical orthogonal functions (EOFs) and variational methods.

The computing resources at the Centre had also been upgraded at regular intervals, replacing the first Cray XMP-14 with Cray XMP-216, acquiring PARAM-1000 computer from CDAC, Pune and the Cray SVI system. At present, NCMRWF produces 7-day predictions using a global T254L64 model; 8-member ensemble 5-day predictions with a global T80L18 model; and 3-day predictions with limited area models (i.e.) the ARPS, MM5, WRF, ETA models at different horizontal resolutions for high impact weather forecasting. Global and regional data assimilations are made using 3-D Variational (3DVAR) assimilation method.

IIT-Delhi: Efforts on development of atmospheric models and numerical weather prediction were initiated at IIT-Delhi in 1979 through the establishment of the Centre for Atmospheric Science (CAS) with the support of IMD and the Ministry of Education. The Centre used Indo-French and Indo-UK Programs to enhance their atmospheric modeling efforts. The first big effort was the import of French global atmospheric model through collaboration with Prof. Sadourny's group at the Laboratoire de Météorologie Dynamique (LMD), France. Similarly, a storm surge model from collaboration with Prof. B. Johns of the University of Reading, UK and a regional weather prediction model of the US Naval Research Laboratory (NRL) were acquired for research. In later years, mesoscale models such as MM5, ARPS and WRF were extensively used for simulation and prediction of mesoscale atmospheric phenomena over India. Along with, a global T21 model from ECMWF was imported and used for long-range forecasting of Indian monsoon. With several of the research programs related to atmospheric modeling, the CAS of IIT-Delhi had become one of the major research Centres by 1985 and contributed to the progress of NWP in India.

Andhra University: The Department of Meteorology and Oceanography is the first in Asia to have initiated teaching and research in meteorology as early as 1949. Teaching and research in NWP was started in 1977 at this department by Bhaskar Rao. At first, a simple barotropic model was developed for which the computer programs were written locally and run on IBM1130 computer. Many students received education and training in NWP continuously as they contributed to the research through their Master theses. As University Grants Commission (UGC), Government of India started M.Tech.

program in Atmospheric Sciences at Andhra University in 1988 as part of national human resource development which helped extension of NWP education to M.Tech. students. Since 1949, many students received education in NWP at Andhra University, who later joined various national institutions and contributed to NWP in India. Bhaskar Rao received exposure and to atmospheric modeling and NWP, especially related to tropical cyclones, through two visits under INSA-JSPS program. Bhaskar Rao was responsible for developing tropical cyclone models for the North Indian Ocean region. Advancements in NWP continued along with the availability of computer resources. The legacy of education in atmospheric modeling and NWP is continuing and Andhra University is identified as a leading contributor for NWP at the national level.

2. Hierarchy of Atmospheric Models

Numerical weather prediction is a method that uses a set of mathematical equations. describing the behaviour of the atmosphere, which can be solved to predict the future state of the atmosphere. This implies that the evolution of the atmosphere could be expressed in a mathematical form, that consists of a set of the mathematical equations, which are independent and whose number is equal to the number of the atmospheric variables (mathematically denoted as a closed system) indicating that the system of equations could be solved for a unique solution using known analytical or numerical methods.

Here, the term "weather" refers to the description of the state of the atmosphere at a particular time, in terms of the magnitudes of the basic variables (i.e.) temperature, pressure, density, humidity, wind (direction and speed), which vary in space and time continuously. Apart from these six variables, rainfall is an important weather parameter that needs to be reported due to its societal importance and impact. It is well known that rainfall is not continuous in space and time and its occurrence over a region results due to intricate interactions of dynamical and physical processes and so will be discussed later. At this time our attention would be confined to the above noted six variables. Earth's atmosphere is a mixture of gases, denoted as "air", surrounding the spherical earth due to force of gravity. The planetary properties of the earth, (i.e.) its location in space with respect to "Sun", rotating on its own axis completing one revolution in 24-hours and its revolution around Sun in one year are important to explain the dynamical behaviour. Similarly the radiation energy received from Sun will drive the atmosphere thermodynamically. Thus, the equations are to represent the dynamical and physical processes that control the evolution of the atmosphere. The atmosphere has a mass of about 5.15 × 1018 kg, with the property continuously changing through movements of air due to changes in the thermo-dynamical state.

Naturally, this brings forth the application of basic laws of conservation, (i.e.) which are the conservation of mass, momentum and energy for the earth-atmosphere system. Our basic knowledge of dynamics and physics categorically indicate fundamental Newton's laws of motion for conservation of momentum; equation of continuity for conservation of mass and thermodynamic energy equation for conservation of energy.

Specifically, the 2nd law of motion states that "The acceleration of an object as produced by a net force is directly proportional to the magnitude of the net force, in the same direction as the net force, and inversely proportional to the mass of the object"; the continuity equation states that "the rate at which mass enters a system is equal to the rate at which mass leaves the system plus the accumulation of mass within the system"; and the first law of thermodynamics states that "the change in the internal energy of a closed system is equal to the amount of heat supplied to the system, minus the amount of work done by the system on its surroundings".

The first two equations of motion are written as

$$\frac{du}{dt} - fv + \frac{1}{\rho}\frac{\partial p}{\partial x} = 0 \qquad .....(2.1)$$

$$\frac{dv}{dt} + fu + \frac{1}{\rho}\frac{\partial p}{\partial y} = 0 \qquad .....(2.2)$$

where u and v denote the zonal (west-east) and meridional (south-north) components of horizontal wind; f is Coriolis force; p is pressure. In meteorology, the zonal wind component from west to east and the meridional component from south to north are conventionally taken as positive.

These two equations, (2.1) and (2.2) are derived from the Newton's 2nd law of motion where the terms $\frac{du}{dt}$ and $\frac{dv}{dt}$ are acceleration terms; and the other terms fv and fu denote the Coriolis force terms, and the terms $\frac{1}{\rho}\frac{\partial p}{\partial x}$ and $\frac{1}{\rho}\frac{\partial p}{\partial y}$ denote the pressure gradient force terms.

The terms $\frac{du}{dt}$ and $\frac{dv}{dt}$ denote total derivatives, which are equal to the local derivative term and the advection terms. Mathematically,

$$\frac{du}{dt} = \frac{\partial u}{\partial t} + u\frac{\partial u}{\partial x} + v\frac{\partial u}{\partial y} + w\frac{\partial u}{\partial z}$$

Where w is the vertical component of the wind (positive for upward motion). The third equation of motion written for the acceleration of "w", in a simplified form, using scale analysis reduces to a diagnostic form as

$$\frac{\partial p}{\partial z} + g\rho = 0 \qquad \text{.....(2.3)}$$

where p is pressure; g is acceleration due to gravity; and ρ is density.

The equation of mass continuity (conservation of mass) states that mass is conserved when a fluid is in motion and the motion must be in such a way that mass is conserved. The equation could be written as

$$\frac{\partial \rho}{\partial t} + u\frac{\partial \rho}{\partial x} + v\frac{\partial \rho}{\partial y} + w\frac{\partial \rho}{\partial z} + \rho\left(\frac{\partial u}{\partial x} + \frac{\partial v}{\partial y} + \frac{\partial w}{\partial z}\right) = 0 \qquad \text{.....(2.4)}$$

where ρ is density; u, v, w are the zonal, meridional and vertical components of the wind.

In addition to these three basic conservation equations, a few more equations relating atmospheric variables are also needed. They are;

The equation of state is a thermodynamic equation based on the ideal gas law, which states that the product of the pressure and the volume of one gram molecule of an ideal gas is equal to the product of the absolute temperature of the gas and the universal gas constant. Mathematically, the equation is

$$Pa = RT \text{ or } P = R\rho T \qquad \text{.....(2.5)}$$

where p is pressure, α is volume, ρ is density, T is temperature and R is gas constant.

This law leads to the definition of "potential temperature" which has application in meteorology due to its conservation property in adiabatic conditions. The potential temperature is thus defined as "the temperature that an unsaturated parcel of dry air would have, if brought adiabatically (without exchange of heat with surroundings) from its initial state to a standard pressure, p_0, typically 1000 hPa". Its mathematical expression is

$$\frac{d\theta}{dt} = \frac{d}{dt} T(Po / P)^k = 0 \qquad \text{.....(2.6)}$$

$$\theta = T(Po/P)^k = 0 \qquad \text{.....(2.7)}$$

where θ is the potential temperature, *T* is temperature, and *k* is the Poisson constant, which equals to the ratio of the gas constant (R) to the specific heat capacity at constant pressure (c_p) for an ideal gas.

The hydrostatic equation, which states that the air pressure at any height in the atmosphere is due to the force per unit area exerted by the weight of all of the air lying above that height. Consequently, atmospheric pressure decreases with increasing height above the ground. Mathematically, the equation is written as

$$\frac{\partial p}{\partial z} = -g\rho$$

This has large implications for developing the system of weather prediction. This equation implies that the change in pressure with the change in height is equal to the average density of the air times the gravitational constant. The negative sign is due to the fact that pressure decreases with height. Density of air, the variable appearing on right hand side, is a function of temperature and moisture content. Increasing water vapor content and/or increasing the temperature causes the density to decrease. In very cold air, the air is very dense. Therefore, the dP/dz is large in cold air (the change in pressure with height is large), meaning that pressure decreases rapidly in cold air and the thickness of a cold air mass is small (the atmosphere is thinner in the vertical when it is cold). In warm air opposite is the case (i.e.) warm air expands and takes up a larger volume. The thickness of warm air and the depth of the atmosphere is greater in a warmer air mass. Since warm air has a low density, the change in pressure with height in warm air is smaller than in cold air.

Although humidity (the presence of water vapour) is an important atmospheric variable, its continuity equation is not included at this stage to make the understanding simpler as confined to dry atmosphere. A representative equation will be added at a later stage.

The above description denotes seven equations [(2.1) – (2.7)], involving seven variables: u, v, w, P, ρ, T and θ; and denote a closed system. However the equations are not in a form that could be solvable directly. Specifically, the equations do not provide explicit calculation of pressure 'p' and vertical velocity 'w'. So, there is need for manipulations of the available set of equations transforming them so as to make them suitable for direct application.

As a first step, equations (2.3), (2.5) and (2.7) could be combined to arrive at an equation suitable for the derivation of pressure,

$$\left(\frac{p}{po}\right)^{k} = -\frac{gk}{R}\int\frac{1}{\theta(z)}\partial z \qquad \text{.....(2.8)}$$

This equation facilitates the computation of pressure at different levels from the vertical distribution of potential temperature.

As second step, the equation of continuity is modified by combining with the thermodynamic energy equation, leading to

$$w = -\int_{0}^{Z}(\nabla.V)dz + \frac{1}{\gamma}\int_{0}^{Z}\left(\frac{1}{p}\right)\int_{z}^{\infty}\left(\frac{\partial V}{\partial z}.\nabla p - \frac{\partial P}{\partial Z}\nabla.V\right)dz\,'\,z'' \qquad \text{.....(2.9)}$$

This equation facilitates the computation of vertical velocity 'w' from the horizontal divergence, and terms involving vertical variation of wind and pressure.

Finally we consider the equations (2.1), (2.2) and (2.6) which are the three prognostic equations for u, v, and θ and the equations – (2.6), (2.7), (2.8) and (2.9) which are four diagnostic equations involving different variables to discuss the problem of weather prediction.

Starting at a time, where from the forecast is to be generated, the observations of the four variables u, v, T and p are considered available all over the 3- dimensional spatial domain. Initially, as a first step, the values of T and p are used to compute θ using equation (2.7). Similarly, the values of T and p are used to compute "α (or ρ)" at every point using equation (2.5). Now with the distributions of the six variables, "w" is computed using equation (2.9). Thus the spatial distributions of all the seven variables u, v, w, P, ρ, T and θ are obtained at every at the initial time t. Now the forecasting procedure, (i.e.) to obtain future state of the seven variables is presented. The three prognostic equations, (2.1), (2.2) and (2.6) are used to compute the rate of change of the variables u, v and θ and estimates at a future time t+ Δt are obtained through using an appropriate numerical method. Now the diagnostic equations are used to compute the remaining three variables. The estimated vertical distribution of θ is used to compute the pressure "p" values in the vertical using equation (2.8). Then, T values are calculated using the distributions of p and θ. As the next step, p and T values are used to compute ρ at every space point. Finally, the values of "w" are computed using the necessary variables in equation (2.9). This completes the estimation of all variables corresponding to a future time t+ Δt. Now the procedure is repeated till the forecast period is complete.

The above derivation of the system of equations and description of the computational procedure enunciates the dynamical weather prediction as a mathematical process. Now the implementation of this system for weather prediction in practical terms will be examined. Although the system is mathematically flawless, its application for weather prediction in the earth-atmosphere system needs to be ascertained. The first important characteristic of the prognostic equations is that they are nonlinear due to the presence of the advection terms. As such this prohibits their direct solution, and leads to the use of numerical methods for their solution. Apparently, suitable finite difference and integration methods are to be used for the calculation of differentials and integrals. Another constraint is of the boundary conditions, which prohibit mass flow across the top of the atmosphere and through the earth surface. The computation of "θ" and "w" at the initial time using equations [(2.8) and (2.9)] reveals that the integrations are to be made up to infinite heights, and a practical limitation for this computation is that the observations could be made up to finite heights only. Both the integrals and differentials which are approximated through numerical methods require spatial observations as densely as possible and also as accurately as possible. This implies that the accuracy of the numerical approximations of the derivatives and integrals would be dependent on the resolution (i.e.) higher accuracy with higher resolution.

Another important issue to be noted is the computation of accelerations using the equations (2.1) and (2.2), in which the magnitudes of the Coriolis force and pressure gradient force terms are an order higher than all other terms. This implies that the computed acceleration is an order smaller than the force terms, and that the accuracy in the observations of wind and pressure are important. Supposing that the wind and pressure are observed with 1% error, then the computed accelerations will have an error of 10% and if the errors were to be 10% then the computed accelerations will have 100% errors and their use in subsequent calculations will lead to spurious forecasts as the errors compound with the every time step of integration. In fact, the errors in the observations will manifest as spurious high frequency oscillations which will grow in amplitude ultimately maskimg the meteorologically important low frequency atmospheric variations. Another important issue is the calculation of vertical velocities, which is primarily dependent on the magnitude of the divergence. A careful assessment of the terms involved in the computation of divergence denote that the terms $\partial u/\partial x$ and $\partial v/\partial y$ are of equal magnitude but of opposite sign, and so the resultant of the sum of $\partial u/\partial x$ and $\partial v/\partial y$ will be an order of magnitude smaller than either of the two terms. Following the same analogy, it is easy to realise that the errors in the observations of u and v are crucial in determining the divergence and further the vertical velocity. If the errors in the observations of u and v are to be 1% (10%), then the errors in divergence would be 10% (100%). As the errors in the wind observations are nearly 10%, the computations of divergence and vertical velocity would be having errors of 100%. So the calculation of the accelerations and the vertical velocity using the observations of wind would pose a problem for application of this procedure. These are also the reasons for the failure of Richardson's numerical weather prediction experiment. The above arguments lead us to conclude that the atmosphere preserves a state of balance which could not be manifested in the observations. If the mathematical system of equations along with numerical methods for the solution is to be used, it may become unavoidable to impose some conditions of balance while defining the initial state of the atmosphere using observations. The preceding discussion leads us to identify the balance properties of the mass and wind fields for their use in the development and application of atmospheric models for weather prediction.

In this chapter, a mathematical system consisting of equations governing the atmospheric motion suitable for weather prediction has been presented. The equations are partial differential equations with nonlinear terms, which necessitate the use of numerical methods for their solution. The problem of weather prediction is referred to as "numerical weather prediction" although the formulation is mathematical.
In chapter 3, the numerical methods that are used to solve the weather prediction system are presented.

2.3 Finite Difference Method

In this method a differential is expressed in terms of finite differences following its definition as

$$\frac{dA}{dX} = \lim_{\Delta x \to 0} \frac{\Delta A}{\Delta x}$$

This implies that the difference of the variable 'A' (i.e.) "ΔA" for a small measure of "Δx" tend to define the rate of change of the variable 'A' over the span of "Δx". Since the value of Δx is to be finite and small, the method is called as finite difference method and $\frac{\Delta A}{\Delta x}$ is the finite difference approximation of $\frac{dA}{dX}$. Similar analogy could be assumed for all space and time dimensions in the case of solution of the partial differential equations of the weather prediction system.

More details of the finite difference methods and their properties are given in chapter 3.

Considering the applicability of the finite difference method, certain issues exist. The derivation of the distributions of "p" and "w" using equations 2.8 and 2.9 involve vertical integrations up to infinite heights, and even if the top of the atmosphere is considered as upper limit of integration, observations are not available to the vertical heights needed such as 10 hPa pressure level. Restriction of the vertical integrations to the top of tropopause indicates errors of approximation.

Another important limitation is the computation of acceleration terms using equations 2.1 and 2.2. Scale analysis of these two equations show that the magnitude of the pressure gradient and Coriolis force terms are of equal magnitude and are at least one order or more higher than the other terms of acceleration, advection and friction. This means that the pressure gradient force and Coriolis force terms are of equal order of magnitude but with opposing signs, so nearly balance each other leading to the computation of resultant acceleration. In this context, if observations of u, v and p have 1% error, the computed acceleration term will have 10% as error. Since the observations of u, v and p have errors higher than 1%, and often as 10%, the errors in the computation of horizontal acceleration would be 100%. Similarly the computation of horizontal divergence, given as $\frac{\partial u}{\partial x}+\frac{\partial v}{\partial y}$ will have unacceptable limits of error as the scale analysis shows the magnitude of horizontal divergence is one order smaller than either of the two terms $\frac{\partial u}{\partial x}$ and $\frac{\partial v}{\partial y}$. An error of 10% in the observation implies an error of 100% in the estimation of horizontal divergence, which is dominant in the computation of the distributions of pressure and vertical velocity through the use of equations 2.8 and 2.9.

The above analysis implies that although the observed large scale atmosphere is in a near balance, the derived atmospheric flow using the mathematical system of equations (2.1) to (2.9) and the finite difference method will show spurious accelerations indicating the presence of high frequency fluctuations. This means that the real atmosphere is in a balanced state than evinced from observations (due to instrument errors). Therefore it is of importance to ascertain the balanced state of the atmosphere and mathematically impose on the system towards possibly evolving a slightly different set of equations which would dynamically evolve the atmosphere nearer to the true state. Since the observations of pressure and horizontal wind are the constituents of the balance, the possible variations of the balance between wind and pressure will bring forth different balance conditions. This will be further elaborated in the next sections of this chapter.

2.3.1 Filtered Models

It is common to perform scale analysis to compare the magnitudes of the governing equations for the atmosphere following Charney (1948).

In this method, the atmospheric variables are assumed to have characteristic values as follows

L = horizontal length scale (~ ¼ wavelength) ~ 10^6 m (1000 km)

D = Vertical scale ~ 10^4 m (10 km)

H = Scale height (i.e.) height of tropopause ~ 10^4 m (10 km)

V = horizontal velocity ~ 10 m/sec

W = Vertical velocity ~ 10^{-2} m/sec

T = Time period = L/V ~ 10^5 sec (27.8 h)

f = Coriolis force ~ 10^{-4} /sec

Further the derivatives are assumed to be

$$\frac{\partial u}{\partial x} \sim \frac{\partial v}{\partial x} \sim \frac{\partial u}{\partial y} \sim \frac{\partial v}{\partial y} \sim \frac{V}{L}$$

$$\frac{\partial u}{\partial z} \sim \frac{\partial v}{\partial z} \sim \frac{V}{D}$$

$$\frac{\partial u}{\partial t} \sim \frac{V}{L/V} \sim \frac{V^2}{L}$$

$$\beta = \frac{\partial f}{\partial y} \sim 10^{-11}\ \text{m}^{-1}\ \text{sec}^{-1}$$

$$\rho = \text{density} = 1\ \text{kg m}^{-3}$$

$$\frac{\Delta\rho}{\rho} \sim 10^{-2}$$

In general, the scale analysis would be performed on the equations of motion, equation of continuity, thermodynamic energy equation etc to assess the relative magnitudes of each of the terms in order to understand their contributions and to understand the characteristics of the atmospheric motion at different scales.

Here, scale analysis of the vorticity and divergence equations are only considered as relevant to the development of atmospheric models.

It is known that the equations for atmospheric vorticity and divergence are derived combining the two equations of motion for horizontal acceleration (i.e.) u and v.

The vorticity equation is written as

$$\frac{\partial \varsigma}{\partial t} + u\frac{\partial \varsigma}{\partial x} + v\frac{\partial \varsigma}{\partial y} + w\frac{\partial \varsigma}{\partial z}$$

$$= -(f + \varsigma)\left(\frac{\partial u}{\partial x} + \frac{\partial v}{\partial y}\right) - \left(\frac{\partial w}{\partial x}\frac{\partial v}{\partial z} - \frac{\partial w}{\partial y}\frac{\partial u}{\partial z}\right)$$

$$+ \frac{1}{\rho^2}\left(\frac{\partial \rho}{\partial x}\frac{\partial p}{\partial y} - \frac{\partial \rho}{\partial y}\frac{\partial p}{\partial x}\right)$$

$$\varsigma = \frac{\partial v}{\partial x} - \frac{\partial u}{\partial y} \sim \frac{V}{L} \sim \frac{10ms^{-1}}{10^6 m} = 10^{-5}\text{s}^{-1}$$

$$\text{f} = 10^{-4}\text{s}^{-1}$$

For synoptic scale motions, the ratio of relative vorticity ζ to earth vorticity f is

$$\frac{\zeta}{f} = \frac{10^{-5}s^{-1}}{10^{-4}s^{-1}} = 10^{-1} = 0.1$$

This ratio ζ/f is also called Rossby number, R_o which is ~ 0.1 for synoptic scale motions

$$\frac{\partial\zeta}{\partial t} \sim u\frac{\partial\zeta}{\partial x} \sim v\frac{\partial\zeta}{\partial y} \sim \frac{V^2}{L^2} \sim \frac{(10\ ms^{-1})^2}{(10^6 m)^2} \sim 10^{-10}s^{-2}$$

$$w\frac{\partial\zeta}{\partial z} : \frac{w\,V}{L\,H} : \frac{(10^{-2}ms^{-1})(10ms^{-1})}{(10^6 m)(10^4 m)} : 10^{-11}s^{-2}$$

$$v\frac{\partial f}{\partial y} \sim V\beta \sim (10^{-11}m^{-1}s^{-1})(10\ ms^{-1}) \sim 10^{-10}s^{-2}$$

$$\frac{\partial u}{\partial x} \sim \frac{\partial v}{\partial y} \sim 10^{-5}s^{-1}$$

Since $\frac{\partial u}{\partial x}$ and $\frac{\partial v}{\partial y}$ are of equal magnitude but of opposite sign

$$\frac{\partial u}{\partial x} + \frac{\partial v}{\partial y} \sim 10^{-6}s^{-1}$$

$$f\left(\frac{\partial u}{\partial x} + \frac{\partial v}{\partial y}\right) \sim 10^{-10}s^{-2}$$

$$\zeta\left(\frac{\partial u}{\partial x} + \frac{\partial v}{\partial y}\right) \sim 10^{-11}s^{-2}$$

$$\left|(f+\zeta)\left(\frac{\partial u}{\partial x} + \frac{\partial v}{\partial y}\right)\right| \sim 10^{-10}s^{-2}$$

Tilting term

$$\frac{\partial w}{\partial x}\frac{\partial v}{\partial z} - \frac{\partial w}{\partial y}\frac{\partial u}{\partial z} \sim \frac{wV}{LH} \sim \frac{(10^{-2}ms^{-1})(10ms^{-1})}{(10^6 m)(10^4 m)} \sim 10^{-11}s^{-2}$$

Solenoid term

$$\frac{1}{\rho^2}\left(\frac{\partial \rho}{\partial x}\frac{\partial p}{\partial y}-\frac{\partial \rho}{\partial y}\frac{\partial p}{\partial x}\right)\sim\frac{1}{\rho}\frac{\Delta \rho}{\rho L}\sim 10^{-11}s^{-2}$$

$$\left(1\text{Pa} = 1\ \text{kg}\ m^{-1}s^{-2} = 1\text{N}\ m^{-2}\right)$$

A hierarchy of atmospheric models were developed based on filtering approximation. In the previous chapter, the need for use of finite difference methods for the solution of atmospheric governing equations for weather prediction has been illustrated. There are several numerical methods that are applied to the weather prediction system of equations, of which finite difference method is the most used. However the properties of the finite difference methods are to be studied to understand the behaviour of the numerical solutions from the use of finite difference methods. These are described in chapter 3.

In brief, a comparison of the three different methods (i.e.) "Forward finite difference method", "Centered finite difference method" and "Implicit method" show that the numerical solutions from these three methods are widely different. While the forward finite difference method leads to amplifying solutions always, the centered finite difference method produce conditionally stable solutions i.e. non-amplifying solutions as long as Courant-Friedrichs-Lewy (CFL) criterion is satisfied; and the "Implicit method" providing stable solutions always without any conditions. It is noted that small differences in the way of approximation of the derivatives lead to substantiated differences in their solutions. Although "Implicit method" is to be preferred, the method is not computationally viable due to implied solution of simultaneous equations of an order equal to the total number of grid points of the domain. Barring the use of forward finite difference method due to its property of leading to unstable solutions, centered finite difference method is the only suitable application although subject to the condition of CFL criterion.

The CFL condition states that $c\frac{\Delta y}{\Delta x} <= 1$ where c is the phase speed of the fastest waves present in the atmosphere model solutions, Δt is the time step at which the rate of change would be computed and Δx is the grid spacing for obtaining the finite differences.

It is known that the atmospheric governing equations have solutions for several wave types and basically for the three known types of longitudinal waves (sound waves); vertical transverse waves (gravity waves) and the horizontal transverse waves (Rossby waves). The longitudinal waves occur due to variations in the (temperature) pressure arising from compression or rarefaction and propagate in all directions with a phase speed of $\bar{u} \pm \sqrt{\gamma RT}$ ($\gamma = c_p / c_v$, R= gas constant and T is the temperature) meaning to be dependent on the physical properties of the medium. The gravity waves have the physical mechanism of the density (pressure) variations arising due to variations of air column at any point and have the particle movement in vertical direction as the waves horizontally propagate with a phase speed given by $\bar{u} \pm [gH(1 - \frac{\rho_1}{\rho_2})]^{1\backslash 2}$,where $\bar{u}$ is mean horizontal wind, g is the acceleration due to gravity; H is depth of bottom fluid; ρ_1 and ρ_2 are densities of upper and bottom fluids. For the gravity waves at the atmosphere-ocean interface, the formula could be simplified as $c = u \pm \sqrt{gH}$ (density $\rho_1 << \rho_2$).
In contrast, Rossby waves have the particles moving in the horizontal plane perpendicular to the propagating horizontal direction with the speed given by $c = \bar{u} - (\beta L^2 / 4\pi^2)$ where $\bar{u}$ is the mean horizontal wind , β is the Rossby parameter $(\frac{\partial f}{\partial y})$ and L is the wavelength.

A comparison of the phase speeds denotes that Rossby waves are the slowest with approximate speed of 10 m/sec (with horizontal wavelengths of 5000 km); gravity wave speeds range from 30 to 300 km/hr; and the sound waves have speeds of ~1000 km/hr.

Atmospheric observations show that the significant weather changes are associated with considerable amplitude of the pressure variations of the order of 10 hPa or more (as compared to daily weather changes are due to pressure variations of 1 or 2 hPa in a day. A comparison of the pressure amplitudes associated with the three wave types show that the sound waves have insignificant pressure variations less than 10^{-1} hPa; gravity waves have pressure amplitudes of 0.1 to 1 hPa; and the zonally propagating Rossby waves have pressure amplitudes of > 10 hPa.

The sound waves arise due to small adiabatic compressibility of air, the gravity waves arise due to incompressibility and vertical density variations (baroclinic), whereas Rossby waves arise due to incompressibility and variations of Coriolis force with latitude (barotropic).

Thus it can be inferred that most of the atmospheric variations leading to significant weather changes are large and of synoptic scale due to slow moving and zonally transient large amplitude waves of the Rossby type and not due to either gravity or sound waves. It is important to note that this reasoning holds good for synoptic scale weather and where meso- and micro-scale weather is to be studied, gravity waves should also be considered.

So, for the synoptic scale weather prediction, the equations of the weather prediction system could be modified by eliminating the numerical solutions corresponding to the sound and gravity waves and at the same time not altering the solutions corresponding to Rossby waves.

This procedure is called “filtering” and the atmospheric models based on the filtering of gravity and sound waves are called “filtered models”. Following the narration of the physical mechanisms responsible for sound and gravity waves, and using the information that adiabatic compression/ rarefaction cause sound waves and hydrostatic pressure changes cause gravity waves, it is mathematically obtained and physically explained that “hydrostatic approximation” leads to the elimination of sound waves and use of “geostrophic approximation” eliminates gravity waves.

Although adiabatic condition has been adopted in the derivation of weather prediction system of equations, it is used only to predict the potential temperature (using equation 2.6), and that the vertical distribution of potential temperature is used to obtain the vertical distribution of pressure (using equation 2.8), which has been obtained using hydrostatic approximation. A detailed mathematical analysis of a mixed gravity-Rossby wave system shows that "geostrophic approximation" eliminates gravity waves.

These two assumptions form the basis of the development of a hierarchy of filtered atmospheric models. Further, use of hydrostatic assumption helps the governing equations to be transformed to pressure as vertical coordinate which simplifies the governing equations to be without explicit presence of density.

Using Helmholtz theorem, treating the horizontal wind to be sum of its rotational and divergent parts.

$$V = V_\psi + V_\chi$$

where $$V_\psi = k \times \nabla\psi \text{ and } V\chi = \nabla\chi$$

Ψ is the stream function associated with the rotational part and χ is velocity potential associated with the divergent part. The mathematical form of their relation is expressed as

$$\zeta = \Delta^2\psi \text{ and } \delta = \Delta^2\chi$$

Comparing the magnitudes, V_ψ is of the same order as total wind whereas V_χ is an order of magnitude smaller than V_ψ.

While formulating a filtered model using vorticity and divergence equations, any simplification through neglect of terms using scale analysis should be affected by imposing the same condition on both of these equations.

The vorticity and divergence equations in p-coordinates are

$$\frac{\partial\zeta}{\partial t} + \left(V_\psi + V_\chi\right).\nabla(\mathrm{f} + \zeta) + \omega\frac{\partial\zeta}{\partial \mathrm{p}} + (\mathrm{f} + \zeta)\nabla.(V_\psi + V_\chi)$$

$$+ \kappa.\nabla\omega * \left(\frac{\partial \mathrm{V}_\psi}{\partial \mathrm{p}} + \frac{\partial \mathrm{V}_\chi}{\partial \mathrm{p}}\right) = 0$$

$$\frac{\partial \delta}{\partial t} + \left(V_\psi + V_\chi\right).\nabla\delta + \omega\frac{\partial \delta}{\partial p} + \delta^2 + \nabla\omega.\left(\frac{\partial V_\psi}{\partial p} + \frac{\partial V_\chi}{\partial p}\right)$$
$$- f\zeta + \beta u - 2J(u,v) + \nabla^2\phi = 0$$

Scale analysis

Vorticity	Divergence
$f \sim 10^{-4}\,sec^{-1}$; $\frac{V}{L} \sim 10^{-5}$ $(R_i R_1)^{-1} \sim 10^{-1}$ $R_0 \sim 10^{-1}$ $R_1 = R_0 =\sim 10^{-1}$ $L = 10^3\,km$ $H = 10\ km$ $V = 10\ m.s^{-1}$	

Vorticity	Divergence
$\frac{\partial \zeta}{\partial t} = \frac{V^2}{L^2} \sim 10^{-10}\,s^{-2}$	$\frac{\partial \delta}{\partial t} \sim 10^{-12}\,s^{-2}$
$V_\psi.\nabla\zeta = \frac{V^2}{L^2} \sim 10^{-10}\,s^{-2}$	$V_\psi.\nabla\delta \sim 10^{-11}\,s^{-2}$
$V_\chi.\nabla\zeta = \frac{V^2}{L^2} \sim 10^{-11}\,s^{-2}$	$V_\chi.\nabla\delta \sim 10^{-12}\,s^{-2}$
$V_\psi.\nabla f = 10^{-10}\,s^{-2}$	$\omega\frac{\partial \delta}{\partial p} \sim 10^{-12}\,s^{-2}$
$V_\chi.\nabla f = 10^{-10}\,s^{-2}$	$\delta^2 = 10^{-12}\,s^{-2}$
$V_\chi.\nabla(f+\zeta) \sim 10^{-11}\,s^{-2}$	$f\zeta = 10^{-9}\,s^{-2}$
$V_\psi.\nabla(f+\zeta) \sim 10^{-10}\,s^{-2}$	$J(u,v) = 10^{-10}\,s^{-2}$
$\omega\frac{\partial \zeta}{\partial p} \sim 10^{-11}\,s^{-2}$	$\beta u = 10^{-10}\,s^{-2}$
$(f+\zeta)\nabla.V_\psi \sim 10^{-10}\,s^{-2}$	$\nabla^2\phi \sim 10^{-9}\,s^{-2}$
$(f+\zeta)\nabla.V_\chi \sim 10^{-11}\,s^{-2}$	$\nabla\omega.\frac{\partial V_\psi}{\partial p} \sim 10^{-11}\,s^{-2}$
$K.\nabla\omega \times \frac{\partial V_\psi}{\partial p} \sim 10^{-11}\,s^{-2}$	$\nabla\omega.\frac{\partial V_\chi}{\partial p} \sim 10^{-12}\,s^{-2}$
$K.\nabla\omega \times \frac{\partial V_\chi}{\partial p} \sim 10^{-12}\,s^{-2}$	

From scale analysis and applying energy constraints we can deduce the following pairs of equations

Pair 1: This pair of equations is called non-linear balanced system

$$\frac{\partial \zeta}{\partial t} + V.\nabla(f+\zeta) + \omega\frac{\partial \zeta}{\partial p} + (f+\zeta)\nabla.V_{\chi} + k.\nabla\omega \times \frac{\partial V_{\psi}}{\partial p} = 0$$

$$\nabla^2\phi + k \times V.\nabla f - f\zeta - 2J(u_{\psi}, v_{\psi}) = 0$$

Pair 2: This pair is called linear balance system

$$\frac{\partial \zeta}{\partial t} + V_{\psi}.\nabla(f+\zeta) + V_{\chi}.\nabla f + f\nabla.V_{\chi} = 0$$

$$\nabla^2\phi + \nabla.(fk + V_{\psi}) = 0$$

or

$$\frac{\partial \zeta}{\partial t} + V_{\psi}.\nabla(f+\zeta) + V_{\chi}.\nabla f = f\frac{\partial \omega}{\partial p}$$

$$\nabla^2\phi - \nabla.(f\nabla\psi) = 0$$

Pair 3: This pair is called quasi-geostrophic system

$$\frac{\partial \zeta}{\partial t} + V_{\psi}.\nabla(f+\zeta) = \bar{f}\frac{\partial \omega}{\partial p}$$

$$\nabla^2\phi - \bar{f}\nabla^2\psi = 0$$

The quasi-geostrophic system can further be simplified by considering barotropic atmosphere and the equation will be

$$\frac{\partial \zeta}{\partial t} + V_{\psi}.\nabla(f+\zeta) = 0$$

$$V_{\psi} = k \times \nabla\psi = \frac{1}{\bar{f}} k \times \nabla\phi$$

which is same as

$$\nabla^2\phi - \bar{f}\nabla^2\psi = 0$$

Now starting from the simplest of the approximations, the development of a hierarchy of models will be presented.

2.4 Barotropic Model

The barotropic vorticity equation is $\frac{\partial \zeta}{\partial t} + u\frac{\partial \zeta}{\partial x} + v\frac{\partial \zeta}{\partial y} + \beta v = 0$

$$v = \frac{g}{f}\frac{\partial z}{\partial x}; \qquad u = -\frac{g}{f}\frac{\partial z}{\partial y} \qquad \text{[geostrophic assumption]}$$

$$\zeta = \frac{\partial v}{\partial x} - \frac{\partial u}{\partial y} = \frac{g}{f}\nabla^2 z$$

$$\frac{\partial}{\partial t}(\frac{g}{f}\nabla^2 z) - \frac{g^2}{f^2}\frac{\partial z}{\partial y}\frac{\partial}{\partial x}(\nabla^2 z) + \frac{g^2}{f^2}\frac{\partial z}{\partial x}\frac{\partial}{\partial y}(\nabla^2 z) + \frac{g}{f}\frac{\partial z}{\partial x}\beta = 0$$

Eliminating g/f throughout from left hand side,

$$\frac{\partial}{\partial t}(\nabla^2 z) + \frac{g}{f}\left[\frac{\partial z}{\partial x}\frac{\partial}{\partial y}(\nabla^2 z) - \frac{\partial z}{\partial y}\frac{\partial}{\partial x}(\nabla^2 z)\right] + \beta\frac{\partial z}{\partial x} = 0$$

This equation involves only one variable 'z' and can be solved numerically using an appropriate method. Since the equation is Helmholtz type of equation, relaxation methods are used for its solution. The above equation, solved for only one variable, is applied at one level, which is the level of non-divergence. So while applying this simple model, the prediction should be made for the level of non-divergence and variations at this level apply at all levels in the vertical. This is because of the assumption of barotropy, which means that density is a function of pressure alone (not of temperature) which also means that the pressure, density and temperature surfaces are parallel.

2.5 Equivalent Barotropic Model

Since the assumption of non-divergent wind implies that changes in relative vorticity can result solely due to its advection, the next refinement of the model would be to include "divergence" in the vorticity equation. However to keep the assumption of barotropic atmosphere, wind direction is considered invariant vertically and vertical variation of wind speed is only considered. With this assumption, wind speed at any level could be expressed in terms of a function derived from the vertical variation of wind speed as

$V = A(p)\overline{V}$ where $\overline{(\,)}$ operator is defined as the vertically integrated value of a variable. Avoiding the mathematical part, the final prediction equation is

$$\frac{\partial \zeta^*}{\partial t} + V^*.\nabla(f + \zeta^*) = \frac{\bar{f}\rho_0 A_0}{p_0}\frac{\partial \bar{\phi}^*}{\partial t}$$

* denotes the level at which $A(P^*) = \overline{A^2}$. This formulation means that the vorticity equation is to be applied at the level where $A(p^*) = \overline{A^2}$, also denoted as "equivalent barotropic level".

As of the simplified model, the variables v and ζ could be expressed in terms of ϕ(or z) and the equation reduces to one equation with one variable as ϕ (or z). The model is called "equivalent barotropic model" because of its application at "equivalent barotropic level". The equation could be solved using "relaxation method" for the "Laplacian equation" form.

As seen from the development of barotropic models, it can be easily understood that the absolute vorticity is nearly conserved (except for the divergence term). Although relative vorticity distributions may be changing, development and decay of pressure systems are constrained due to non-representation of temperature advection (due to coincidence of isobars and isotherms).

The next advancement would be the introduction of horizontal temperature gradients that contributes to horizontal temperature advection. As known from atmospheric dynamics, horizontal temperature gradients give rise to thermal wind which means the inclusion of vertical variations of the horizontal wind field. Although several variants are possible, one simple model formulation will be presented here.

2.6 A Two-level Baroclinic Model

The atmosphere is divided into four layers of equal thickness (for e.g.) dividing the pressure thickness from earth surface (1000 hPa) to top of atmosphere (0 hPa) as 1000, 750, 500, 250 and 0 hPa levels.

Figure 2.1 Vertical structure of the two-level model

Applying the vorticity equation at levels 1 and 3

$$\frac{\partial \zeta_1}{\partial t} + V_1.\nabla(f + \zeta_1) = \bar{f}\left(\frac{\partial \omega}{\partial p}\right)_1 = \bar{f}\frac{\omega_2 - \omega_0}{2\Delta p} \qquad .....(2.10)$$

$$\frac{\partial \zeta_3}{\partial t} + V_3.\nabla(f + \zeta_3) = \bar{f}\left(\frac{\partial \omega}{\partial p}\right)_3 = \bar{f}\frac{\omega_4 - \omega_2}{2\Delta p} \qquad .....(2.11)$$

Now, we assume that the vertical velocity at the top of the atmosphere (pressure level, p=0) is equal to zero and also that ω_4 is zero at $p=p_s$ (i.e.) at the lowest pressure surface.

Also, we can write $V = \frac{V_1 + V_3}{2}$ and $V_T = \frac{V_1 - V_3}{2}$, which means that V and V_T represent the mean wind and the wind shear in the layer bound between levels 1 and 3. This also means that V_T represent the thermal wind. Similar analogy will lead to

$$\zeta = \frac{\zeta_1 + \zeta_3}{2} \quad \text{and} \quad \zeta_T = \frac{\zeta_1 - \zeta_3}{2}$$

which are denoted as average vorticity and thermal vorticity in the layer bound between levels 1 and 3.

Substituting for $V_1 = V + V_T$; $V_3 = V - V_T$; $\zeta_1 = \zeta + \zeta_T$ and $\zeta_3 = \zeta - \zeta_T$ in the equations (2.10) and (2.11), we get (after substitution and simplification)

$$\frac{\partial \zeta}{\partial t} + V.\nabla(f + \zeta) + V_T.\nabla\zeta_T = 0 \qquad .....(2.12)$$

$$\frac{\partial \zeta_T}{\partial t} + V.\nabla\zeta_T + V_T.(f + \zeta) = \frac{\bar{f}\omega_2}{2\Delta p} \qquad .....(2.13)$$

The Thermodynamic energy equation

$$C_p \frac{dT}{dt} - \alpha\omega = Q$$

can be modified (through mathematical formulation), to obtain

$$\frac{\partial}{\partial t}\frac{\partial z}{\partial p} + V.\nabla\frac{\partial z}{\partial p} + \sigma w = 0$$

where $\sigma = \frac{1}{\theta}\frac{\partial z}{\partial p}\frac{\partial \theta}{\partial p} = 0$

Substituting the expressions for V, V_T, ζ and ζ_T using geostrophic approximation

$$V = [k \times \nabla z]$$

$$V_T = \frac{g}{f}[k \times \nabla h]$$

$$\zeta = \frac{g}{f}\nabla^2 z$$

$$\zeta_T = \frac{g}{f}\nabla^2 h$$

where $z = \frac{z_1 + z_3}{2}$ and $h = \frac{z_1 - z_3}{2}$

The thermodynamic energy equation

$$\frac{\partial}{\partial t}(\frac{\partial z}{\partial p}) + V.\nabla(\frac{\partial z}{\partial p}) + \gamma\omega = -\frac{R}{g\rho Cp}Q \qquad .....(2.14)$$

where Q is diabatic heating. Considering adiabatic flow where Q = 0, the equation reduces to

$$\frac{\partial}{\partial t}(\frac{\partial z}{\partial p}) + V.\nabla(\frac{\partial z}{\partial p}) + \gamma\omega = 0$$

Applying this equation at level-2, we will have

$$\frac{\partial h}{\partial t} + V.\Delta h - \sigma\omega_2 \Delta p = 0 \qquad .....(2.15)$$

We have three equations (2.13), (2.14) and (2.15) and three unknowns z, h and ω_2. This closed system can now be solved for prediction. Another simplification can also be made for the two-level model. Eliminating ω_2 by combining equations (2.14) and (2.15), we will have the new equation as

$$\frac{\partial \zeta_T}{\partial t} + V.\nabla\zeta_T + V_T.(f + \zeta) - \frac{\bar{f}}{2\Delta p\sigma}\left[\frac{\partial h}{\partial t} + V.\nabla h\right] = 0 \qquad .....(2.16)$$

As could be noted from the model derivation, ω_2 is a variable that is computed. As ω denotes "vertical velocity", its spatial distribution provides means to identify the regions of "upward" and "downward" motions which could be associated with "disturbed" (region with rain) and "undisturbed"(dry region) areas to help in weather prediction.

An equation for 'ω' vertical velocity (at level 2) could be obtained through elimination of time derivative term combining the two equations (2.14) and (2.15).

Applying Laplacian operator ∇^2 on equation (2.16)

$$\nabla^2\left[\frac{\partial h}{\partial t}+V.\nabla \mathrm{h}\right]-\sigma\nabla^2\omega_2=0$$

$$\frac{\partial\nabla^2 h}{\partial t}+\nabla^2(V.\nabla \mathrm{h})-\sigma\nabla^2\omega_2=0$$

$$\frac{\mathrm{g}}{\bar{f}}\frac{\partial\nabla^2 h}{\partial t}+\left[V.\nabla\zeta_{\mathrm{T}}+V_{\mathrm{T}}.(f+\zeta)\right]=\frac{\bar{f}\omega_2}{2\Delta \mathrm{p}}$$

$$\frac{\partial\nabla^2 h}{\partial t}+\frac{\bar{f}}{\mathrm{g}}\left[V.\nabla\zeta_{\mathrm{T}}+V_{\mathrm{T}}.(f+\zeta)\right]=\frac{\bar{f}^2\omega_2}{2\mathrm{g}\Delta \mathrm{p}}$$

Eliminating $\frac{\partial}{\partial t}\nabla^2\mathrm{h}$, we get

$$\sigma\nabla^2\omega_2-\frac{\bar{f}^2\omega_2}{2\mathrm{g}\Delta \mathrm{p}}=\nabla^2(\mathrm{V}.\nabla\mathrm{h})-\frac{\bar{f}}{\mathrm{g}}\left[\mathrm{V}.\nabla\zeta_{\mathrm{T}}+\mathrm{V}_{\mathrm{T}}.(f+\zeta)\right] \quad .....(2.17)$$

This is a Helmholtz type equation, which can be solved using relaxation methods.

The above derivation of a two-level model system, and a comparison of the prediction equations for vorticity in the 1-level barotropic and 2-level baroclinic models brings out an important dynamical relation exhibited through the term $\mathrm{V}_{\mathrm{T}}.\nabla\zeta_{\mathrm{T}}$. An analysis of this term may be as follows. ζ_{T} means thermal vorticity (i.e.) the part of vorticity due to thermal wind. Mathematically, $\zeta_{\mathrm{T}}=\frac{\mathrm{g}}{\mathrm{f}}\nabla^2\mathrm{h}$ where h is the thickness of isobaric levels. Considering the finite difference notation

$$\nabla^2 h=\frac{h_1+h_2+h_3+h_4-4h_0}{4\Delta x^2}$$

Here h_0 is the value of h at the centre point in a " + " type grid notation and h_1, h_2, h_3 and h_4 denote the values at the cardinal points. The value of $\nabla^2\mathrm{h}$, computed as positive value means relatively small value at the centre. Conceptually this means that thermal velocity is higher or positive in regions of lower 'h' which is analogous to "relative vorticity is positive in regions of smaller z or in low pressure regions". Thus positive regions of thermal vorticity are associated with thermal troughs or "cold" regions [as pressure decrease with height is more in cold region as compared to warmer region].

Considering a wave-like temperature structure

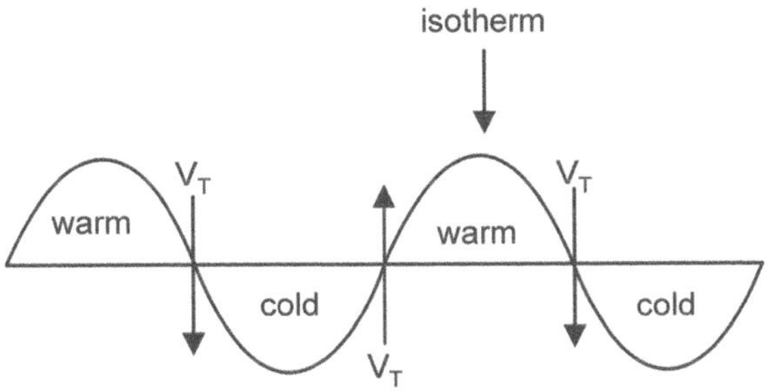

ζ_T is positive in cold regions and negative in warm regions. V_T is the thermal wind, which has direction of motion such as cold regions exist to the left and warm regions are to the right. This means that between the cold and downwind warm regions, V_T is directed as from south to north (or north-south) between the cold and downwind warm (warm and downwind cold) regions. As V_T is positive (negative) in the cold (warm) centres, advection of thermal vorticity by the thermal wind contribute to increase (decrease) of relative vorticity in the region between thermal trough (ridge) and downwind thermal ridge (trough).

This brings out an important synoptic consequence of the passage of pressure systems in relation to spatial thermal structure. If a transient low pressure system will pass over the region between thermal trough and downwind thermal ridge, then the low pressure system will intensify during its passage over the region. Similarly the low pressure system will lose its intensity when it passes over the region between the thermal ridge and the downwind thermal trough.

This analysis shows that development of atmospheric models for weather prediction and their analysis will not only help weather prediction but also to understand dynamical relationships and consequences for changing atmospheric patterns.

2.7 Multi-level Quasi-geostrophic Model

Considering the vertical variation of horizontal atmospheric circulation, a two-level model indicates the simplest formulation of a baroclinic atmosphere and a natural extension would be to design a multi-level model, with the number of vertical levels to be dependent on the application and computational resources.

1. Vorticity Equation

$$\frac{\partial \zeta}{\partial t} + V.\nabla(f+\zeta) = \overline{f}\frac{\partial \omega}{\partial p}$$

$$\frac{g}{f}\nabla^2\frac{\partial z}{\partial t} + V.\nabla(f+\zeta) = \overline{f}\frac{\partial \omega}{\partial p} \qquad .....(2.18)$$

2. Thermodynamic energy equation

$$\frac{\partial}{\partial t}\left(\frac{\partial z}{\partial p}\right) + V.\nabla\left(\frac{\partial z}{\partial p}\right) + \sigma\omega = -Q \qquad ..... (2.19)$$

Q represents diabatic heating, σ is static stability varying with height only.

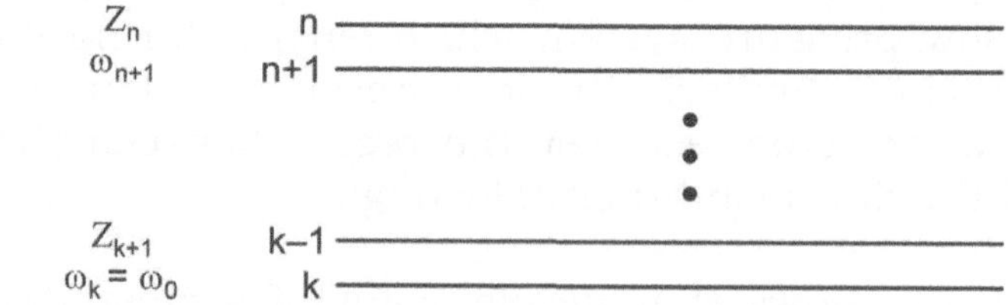

Figure 2.2 Vertical structure of the multi-level model

In this formulation (as shown in figure 2.2), the vorticity equation (height tendency) is applied at odd number levels (i.e.) 1, 3, 5, ...n, ..., k-1 and the ω-equation is applied at even number levels (i.e.) 2, 4, 6, ...n-1, n+1,............k.

The diagnostic equation for ω is obtained by eliminating the time derivative term from equations (2.18) and (2.19). This is achieved by differentiating equation (2.18) with respect to pressure p; and taking the Laplacian (∇^2) of equation (2.19) as follows

$$\frac{\partial}{\partial p}\left(\frac{g}{f}\nabla^2\frac{\partial z}{\partial t}\right)+\frac{\partial}{\partial p}\left(V.\nabla\left(f+\zeta\right)\right)=\overline{f}\frac{\partial^2\omega}{\partial p^2} \quad .....(2.20)$$

$$\nabla^2\frac{\partial}{\partial t}\left(\frac{\partial z}{\partial p}\right)+\nabla^2\left(V.\nabla\frac{\partial Z}{\partial p}\right)+\sigma\nabla^2\omega=Q \quad .....(2.21)$$

Eliminating the time derivative term, we get

$$\sigma\nabla^2\omega+\overline{f}\frac{\partial^2\omega}{\partial p^2}=\frac{\overline{f}}{g}\frac{\partial}{\partial p}\left[V.\nabla\left(f+\zeta\right)\right]-\nabla^2\left[V.\nabla\frac{\partial z}{\partial p}\right]-\nabla^2 Q \quad ....(2.22)$$

Using $\frac{\partial z}{\partial p}=-\frac{1}{g\rho}=\frac{RT}{gp}$

$$\sigma\nabla^2\omega+\overline{f}\frac{\partial^2\omega}{\partial p^2}=\frac{\overline{f}}{g}\frac{\partial}{\partial p}\left[V.\nabla\left(f+\zeta\right)\right]+\frac{R}{gp}\nabla^2\left[V.\nabla T\right]-\nabla^2 Q \quad .....(2.23)$$

This is a diagnostic equation for vertical velocity "ω", which is applied at even number levels. Analysis of the equation (2.23) will indicate how the first two terms on the right hand side [(i.e.) vertical variation of horizontal vorticity advection and the Laplacian of horizontal temperature advection] and the Laplacian of diabatic heating contribute to "ω" vertical velocity.
LHS: - $\nabla^2\omega$ indicates Laplacian of ω; positive values of $\nabla^2\omega$ means smaller values of ω at the centre meaning upward motion.

$\nabla^2\left(V.\nabla T\right)$ means Laplacian of temperature advection.

$V.\nabla T$ is negative means warm air advection.

If $\nabla^2\left(V.\nabla T\right)$ is positive, it contributes to upward motion.

$\frac{\partial}{\partial p}[V.\nabla(f+\zeta)]$ is vertical variation of horizontal vorticity advection. If $V.\nabla(f+\zeta)$ denotes advection of higher (positive) values, $V.\nabla(f+\zeta)$ is negative. If negative $V.\nabla(f+\zeta)$ is increasing upward $\frac{\partial}{\partial p}[V.\nabla(f+\zeta)]$ is positive and contribute to $\nabla^2\omega$ as positive and so negative values of ω (i.e.) upward motion.

In the region of westerlies, with a westward tilt of a trough, this term contributes to upward motion. At higher levels, with trough centre displaced to west of its position below, higher relative vorticity would be advected at upper levels in the locality that correspond to trough centre at lower level. That is, upward motion is derived between the trough and downwind ridge. This also corresponds to the relative positions of the thermal and pressure patterns (i.e.) increase of vorticity (cyclonic or positive) when the thermal wave lags the pressure wave by a quarter of wavelength.

2.8 Linear and Nonlinear Non-geostrophic Baroclinic model

The linear balance system is

$$\frac{\partial \zeta}{\partial x} + V\psi.\nabla(f+\zeta) + V\chi.\nabla f + f\nabla.V\chi = 0 \qquad .....(2.24)$$

$$\nabla.(f\nabla\psi) - \nabla^2\phi = 0 \qquad .....(2.25)$$

Avoiding the mathematics, the final equations could be obtained.

The ω- equation is

$$\sigma\nabla^2\omega + f^2\frac{\partial^2\omega}{\partial p^2} = f\frac{\partial}{\partial p}[V\psi.(f+\zeta)] - \nabla f.\nabla\frac{\partial^2\psi}{\partial p\partial t} + f\nabla f.\frac{\partial v\chi}{\partial p} - \nabla^2\left[V.\nabla\frac{\partial\phi}{\partial p}\right] \qquad ..... (2.26)$$

The vorticity equation is used in the form

$$\frac{\partial \zeta}{\partial t} + V\psi \bullet \nabla\zeta + (V\psi + V\chi) \bullet \nabla f = f\frac{\partial \omega}{\partial p} \qquad .....(2.27)$$

The continuity equation is used in its simplified form

$$\nabla.V = \nabla^2 \chi = -\frac{\partial \omega}{\partial p} \qquad .....(2.28)$$

This system due to the presence of $\frac{\partial^2 \psi}{\partial p \partial t}$ term in the "ω" equation has to be solved using iterative procedure as follows

Solve equation (2.25) to get the stream function ψ field from the ϕ field (known)

Solve equation (2.28) to obtain χ-field

Solve the vorticity equation [equation 2.27] using the first estimates of ω and x fields to obtain $\frac{\partial \psi}{\partial t}$

Solve the ω-equation [equation 2.26] with the estimates of $\frac{\partial \psi}{\partial t}$ and χ to obtain second estimates of ω

Repeat the steps 2, 4 and 3 to revise the estimates till the iterative procedure converges.

The above procedure is time consuming, and generally for operational purposes the terms involving χ and $\frac{\partial^2 \psi}{\partial p \partial t}$ in equation (2.26) are omitted and the equations are solved directly to arrive at a prediction.

2.9 Nonlinear Non-geostrophic Baroclinic Model

The pair of vorticity and balance equations from scale analysis can be written as

$$\frac{\partial \zeta}{\partial t} + V.\nabla(f+\zeta) + \omega\frac{\partial \zeta}{\partial p} + (f+\zeta)\nabla.V\chi + \text{k}.\nabla\omega \times \frac{\partial V\psi}{\partial p} = 0 \quad .....(2.29)$$

$$\nabla^2\phi + \text{k} \times V.\nabla f - f\zeta - 2\text{J}(u_\psi, v_\psi) = 0$$

$$\nabla.(f\nabla\psi) + 2\left[\frac{\partial^2\psi\partial^2\psi}{\partial x^2\partial y^2} - (\frac{\partial^2\psi}{\partial x\partial y})^2\right] = \nabla^2\emptyset \quad .....(2.30)$$

and the thermodynamic equation

$$\frac{\partial\theta}{\partial p} + \text{V}.\nabla\theta + \omega\frac{\partial\theta}{\partial p} = 0 \quad .....(2.31)$$

Using mathematics, we get the ω equation as

$$\nabla^2(\sigma\omega) + f^2\frac{\partial^2\omega}{\partial p^2}$$

$$= f\frac{\partial}{\partial p}J(\psi, f+\zeta) + \frac{\alpha}{\theta}\nabla^2 J(\psi,\theta) - 2\frac{\partial^2}{\partial t\partial p}\left[J(\frac{\partial\psi}{\partial x},\frac{\partial\psi}{\partial y})\right] - f\frac{\partial(\zeta\nabla^2\chi)}{\partial \text{p}}$$

$$+f\frac{\partial}{\partial p}\left(\omega\frac{\partial\zeta}{\partial p}\right) + f\frac{\partial}{\partial p}\left(\nabla\omega.\nabla\frac{\partial\psi}{\partial p}\right) - f\frac{\partial}{\partial p}(\nabla\chi.\nabla(f+\zeta))$$

$$-\frac{\alpha}{\theta}\nabla^2(\nabla\chi.\nabla\theta) - \nabla f.\nabla\frac{\partial^2\psi}{\partial t\partial p} \quad .....(2.32)$$

The formulations of the filtered models described so far in the preceding sections indicate the simplicity of the models such as quasi-geostrophic barotropic and baroclinic models and also the complexity of the linear balanced and nonlinear balanced models. As improvements in weather predictions are demanded, it becomes necessary to use the complex models. It has been clearly illustrated, through the derivation of linear and nonlinear models, of the complexities in solving the divergence equation (in diagnostic form) which inhibit the use of filtered models. Essentially the filtering assumption has been used as it facilitates larger time step "Δt" to satisfy the CFL criterion for numerical stability while adopting centered finite difference approximation and for the computational advantages that provide good lead time for weather predictions. However, use of complex filtered models loses their advantage as their solutions need larger computer time for model integrations.

This argument directs us to the use of primitive equations as derived in the first few sections. The formulation and the properties of atmospheric models using the primitive equations is described in a later section of this chapter. At this stage, it is important to note that the filtered models have been formulated using pressure ‘ p‘ as vertical coordinate.

The hydrostatic equation expresses the relationship between height and pressure in a vertical column of the atmosphere. This facilitates use of pressure as a vertical coordinate, and in the p-coordinate system, horizontal pressure gradients are expressed as gradients of geopotential on an isobaric surface. The important advantage is that density does not explicitly appear in the pressure gradient term. This also means that the geostrophic wind computed using horizontal pressure gradient is a function of density whereas the geostrophic wind calculated using geopotential gradient is a function of height. In the p-coordinate system, equation of continuity is simple and linear of u, v and ω. The p-coordinate system has some disadvantages in the model formulation. The lowest p-surface is assumed to be constant in horizontal space and time. The p-surface may also intersect the earth’s surface such as in the regions of elevated topography (mountains).

Primitive equation models

In meteorology, for the purpose of atmospheric model development, where variations in topography are important, a terrain-following vertical coordinate has been proposed by Philips (1957).

In this a variable “sigma σ” is defined as pressure normalized to its surface value. A simple definition is $\sigma = \frac{p}{p_s}$ where p_s is surface pressure.

This denotes $\sigma = 1$ at the surface and $\sigma = 0$ at the top of atmosphere, where $p = 0$

Alternatively $\sigma = \frac{p - p_T}{p_s - p_T}$, where p_T is the pressure at the top of model atmosphere.

It can be easily seen that $\sigma'(\sigma - \text{vertical velocity}) = \frac{d\sigma}{dt}$

$\sigma' = 0$ at $\sigma = 1$ (surface)

The transformation from p-coordinates to σ -coordinate system is presented in Appendix A.

The sigma coordinates have the advantage of representing the lower boundary i.e., at the earth's surface with implied terrain variations. Since this system is terrain following, the horizontal variations are well represented where the terrain variations are wide and smooth. It also provides increasing vertical resolution near ground with desired consistency. This coordinate does not intersect with the ground as of p-coordinate system.

A disadvantage of the sigma coordinates is the computation of pressure gradient force term in the region of irregular terrain, such as steep mountain region. Errors come in because of the estimation of lapse rates between pressure (height) surfaces. Errors in the estimation of pressure gradient force due to the terrain variations, lead to large errors in the horizontal wind near the surface, and these errors near the surface effect the computations of all variables at different vertical levels. It is also possible that the true heights at different points of the mountain region are represented with smoothened values which result as errors in the estimation of surface pressure, temperature and moisture. Similar effects will be noticeable where the mpuntain region exists near the coast leading to false representation of land extending into oceanic region.

2.10 The Problem of Initialization in PE Models

In PE models, where the governing equations involve both the mass and wind fields, use of the observations of wind and geopotential (mass) directly lead to inconsistencies that contribute to amplifying of initial errors during model integration. This has been the preamble for the development of filtered models, wherein the divergence equation in diagnostic form imposes a balance between the wind and mass fields.

Examining equation (2.34) defining the equation of motion, it has been deduced that the Coriolis force and pressure gradient force terms are of the same magnitude and are atleast one order of magnitude higher than the rest of the terms. This means that the acceleration derived from this equation is dependent on the accuracy with which these terms could be computed from the observations. If the error in the observation of wind or geopotential is 10%, the error in the derivation of acceleration using equation (2.34) would be 100%, since the magnitude of acceleration is one order smaller than the Coriolis force and pressure gradient force terms. If the observations of wind and geopotential are used, the observational errors would be represented as gravity wave oscillations of reasonable amplitude in the model atmosphere although they are not existent in the real atmosphere. This type of spurious errors will spoil the model prediction of the atmosphere as the spurious gravity waves exponentially grow and manifest as fast growing large scale atmospheric peturbations.

There are several ways to overcome this problem. A simple approach is to derive either of the mass and wind fields using observations of one field from the other using diagnostic relation between mass and wind fields.

A geostrophic balance implies derivation of wind field from mass field. However, this will lead to large errors in the region of circular or curved isobars due to presence of inertial force. So, derivation of wind field using gradient wind equation will be devoid of large errors. However use of gradient wind equation is not simple as it involves the use of radius of curvature. If the wind at the initial time is assumed to be nondivergent, then the balance equation can be written considering $\nabla . V = 0$ and $V = V_\psi = k \times \nabla \psi$ where ψ is stream function.

$$\nabla^2\left[\phi + \frac{1}{2}(\nabla\psi)^2\right] = \nabla.\left[(f + \nabla^2\psi)\nabla\psi\right]$$

This equation indicates that ϕ is linear whereas ψ is nonlinear. Generally for middle latitudes, ϕ is observed more consistently and so ψ could be derived from ϕ field. In tropics, where wind field observations are more suitable, ϕ field could be derived from ψ field.

2.11 A Generalised Primitive Equation Model

It is possible to have variety of formulations of atmospheric models using primitive equations. Here in, one version is described based on Shuman and Hovermole (1968) which was used in United States Weather Bureau.
The model equations are written as

$$\frac{\partial u}{\partial t}+\sigma'\frac{\partial u}{\partial \sigma}+m\left(u\frac{\partial u}{\partial x}+v\frac{\partial v}{\partial y}+\frac{\partial\, gZ}{\partial x}+c_p\theta\frac{\partial \pi}{\partial x}\right)$$
$$-v\left(f-v\frac{\partial m}{\partial x}+u\frac{\partial m}{\partial y}\right)+Fx=0 \qquad .....(2.33)$$

$$\frac{\partial v}{\partial t}+\sigma''\frac{\partial v}{\partial \sigma}+m\left(u\frac{\partial u}{\partial x}+v\frac{\partial v}{\partial y}+\frac{\partial gZ}{\partial y}+c_p\theta\frac{\partial \pi}{\partial y}\right)$$
$$+u\left(f-v\frac{\partial m}{\partial x}+u\frac{\partial m}{\partial y}\right)+Fx=0 \qquad .....(2.34)$$

$$\frac{\partial\, gz}{\partial \sigma}+c_p\theta\frac{\partial \pi}{\partial \sigma}=0 \qquad .....(2.35)$$

$$\frac{\partial \theta}{\partial t}+\sigma'\frac{\partial \theta}{\partial \sigma}+m\left(n\frac{\partial \theta}{\partial x}+v\frac{\partial \theta}{\partial y}\right)+H=0 \qquad .....(2.36)$$

$$\frac{\partial}{\partial t}\left(\frac{\partial p}{\partial \sigma}\right)+\frac{\partial}{\partial \sigma}\left(\sigma'\frac{\partial p}{\partial \sigma}\right)+$$
$$m\left(\frac{\partial}{\partial x}(u\frac{\partial p}{\partial \sigma})+\frac{\partial}{\partial y}(v\frac{\partial p}{\partial \sigma})\right)-\frac{\partial p}{\partial \sigma}\left(u\frac{\partial m}{\partial x}+v\frac{\partial m}{\partial y}\right)=0 \qquad .....(2.37)$$

where

$$\pi=\left(\frac{p}{p_0}\right)^{R/c_p} \quad (p_0=1000\text{hPa}) \qquad .....(2.38)$$

$$\frac{\partial q}{\partial t}+\sigma'\frac{\partial q}{\partial \sigma}+m\left(u\frac{\partial q}{\partial x}+v\frac{\partial q}{\partial y}\right)+c=0 \qquad .....(2.39)$$

x and y are the two horizontal Cartesian coordinates

m is map projection ratio

z is geopotential height

θ is potential temperature

p is pressure

R is gas constant for dry air

C_p is specific heat of air at constant pressure

f is the Coriolis parameter

q is specific humidity

F_x and F_y are contributing terms due to friction

H is diabatic heating

C is condensation/ evaporation

The vertical coordinate is adopted as

$$\sigma = \frac{p - p_h}{p_L - p_h}$$

Where p_h is the pressure at air upper surface and p_L is the pressure at the lower surface of an atmospheric layer under consideration. In this model configuration, severe levels are chosen which are divided into four domains and each domain is defined separately. The vertical levels are shown in figure 2.3.

The vertical atmosphere is divided into four domains each of which is separately configured as shown in figure 2.3.

In the lowest layer denoted as planetary boundary layer, σ is defined as

$$\sigma_B = \frac{p - p_5}{p^* - p_5}$$

where p_5 is the pressure at the top of the boundary layer; and p^*-p_5 is assumed to be constant (50 hPa) in space and time.

In the second upper layer denoted as troposphere, which has three levels, σ is defined as

$$\sigma_T = \frac{p - p^{**}}{p_5 - p^{**}}$$

($P^{**}$is the pressure at the tropopause)

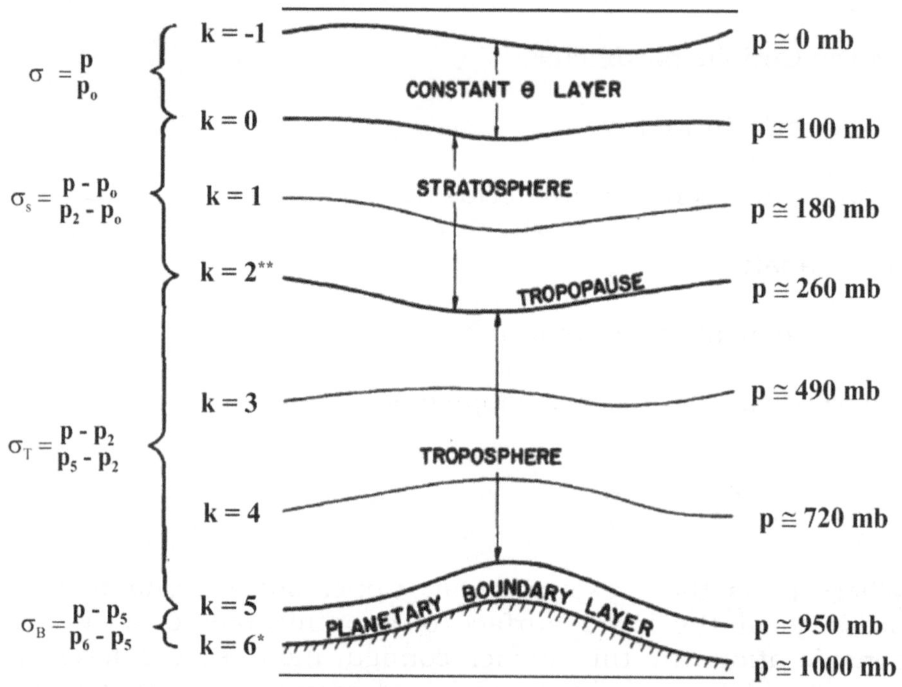

Figure 2.3 Vertical structure of the six-layer premitive equation model

In the third upper layer, denoted as stratosphere, σ is defined as

$$\sigma_s = \frac{p - p_0}{p^{**} - p_0}$$

where p_0 is the pressure at the top of stratosphere.

In the top layer, the domain is defined by

$$\sigma_\theta = \frac{p}{p_0}$$

This top layer is defined solely for computational convenience. Thus the levels 0 and 6 constitute six levels of this primitive equation model.

Model solution

1. The basic inputs of geopotential height and temperature includes values the 10 at standard pressure levels - 1000, 850, 700, 500, 400, 300, 250, 200, 150 and 100 hPa These values are taken from analyses except the temperatures at 1000 hPa and 850 hPa data. These temperature data are used only to determine the lapse rates.

2. Pressures on three reference σ - surfaces (p at 0, p at *level (ground) and p at **level (tropopause) are either analysed or derived. Firstly, the pressure at tropopause is obtained. Secondly, the pressure at the ground is derived using z* (ground elevation), heights of the two closest pressure surfaces and the lapse rate. Thirdly, pressure at tropopause p_0 is specified as some fracftion of the mean tropopause pressure. Alternately p_0 is determined by an intersection of two pressure-height curves, one for the stratosphere and another for the computational layer.

3. The condition that z is constant at p=0 and θ is constant in the computational layer assumes that pressure gradients do not exist initially in the computational layer. With the assumption that θ is constant at initial time, it remains constant due to equation (2.37).

4. Specifications of p*, p** and p_0, along with the boundary layer thickness of 50 hPa as constant provides the initial conditions of $\frac{\partial p}{\partial \sigma}$

5. Pressure at the σ- surfaces are obtained through interpolation from the values at tropopause and constant- pressure data. The interpolation uses both heights and temperatures from isobaric changes but only for height of σ- surfaces.

6. Once the heights of the σ - surfaces are obtained, θ of the layers is determined using equation (2.36).

7. The interpolation for height assumes linear variation of temperature with log P between two isobaric surfaces.

8. The finite difference formulation of this model is given in Appendix C.

3. Numerical Methods

The equations of the weather prediction system narrated in the previous chapter are identified as partial differential equations as they include terms such as $\partial/\partial t$; $\partial/\partial x$; $\partial/\partial y$; $\partial/\partial z$ etc. They are also noted to be nonlinear as the advection terms involve products of two independently varying functions such as u and $\partial/\partial x$. Due to the nonlinearity, the partial differential equations cannot be solved analytically and numerical methods are to be adopted.

The use of finite difference approximations for derivatives as a part of solving the differential equations had been age old, since Euler's application to one dimensional space in 1768 and Runge's extension to two-dimensional space in 1908. The application of finite difference methods was stimulated by the emergence of computers towards solution of intricate problems of science and technology.

The basis of the finite difference method is to approximate the derivatives using differential quotients. At first, the computational domain is to be segregated in space, time steps discretised and then the approximations of the differentials are effected at the space points and for the time step. The error between the numerical solution and the real solution arising due to difference approximation is called as truncation error. As seen later, the truncation error is due to the choice of a finite number of terms out of the Taylor series.

Part A. Finite Differrence Methods

The finite difference method is an approach, in which the derivatives of a function are approximated by the differences of the function values at two successive points of the independent variable separated by a small increment. In this method, the derivative terms in the differential equations are replaced with finite difference approximation in a discretized grid domain. This leads to an algebraic equation for each of the grid point for all the grid points, which can be solved individually at each point (explicitly) or for all points together simultaneously (implicitly) to derive the values of the dependent function using the values of the independent function for the discretized domain.

The derivatives are approximated as

$$\frac{dA}{dx} = \lim_{\Delta x \to 0} \frac{\Delta A}{\Delta x}$$

This means that the differences of the variable A are to be taken between two points of x and smaller the Δx, better is the approximation.

For computational purposes a domain is to be identified. In the case of weather prediction system, the variables are four dimensional as they vary in 3-dimensional space with respect to x, y, and z and the 4th dimension as time.

Since weather prediction comprises earth's atmosphere, suitable formulation is to be made. It is simpler to divide the vertical extent of the atmosphere into a number of layers (i.e.) to have the information represented at discrete verticals levels.

Now that each vertical level is identified, which is a horizontal plane, the domain points in the horizontal plane are to be defined. If x, y axes are chosen for representation of the horizontal plane, the domain could be formulated as the intersection of the horizontal lines in the x and y directions. See figure 3.1.

The x-axis denotes west-east direction and the y-axis represent south-north direction considering the sign of increasing values. As seen from the figure, Δx and Δy are the distances between two successive points in the x- and y- directions. It is to be noted that Δx and Δy need not be the same. The intersecting points are referred to as "grid" points. If there are 'n' points in the x direction and 'm' points in the y-direction, the total number of grid points will be (n x m).

For the representation of grid on the spherical earth, x-values correspond to longitudes and y- values correspond to latitudes. It is to be recognized that Δx values representing the distance between two longitudes, will be decreasing with increase of y due to spherical form of earth whereas the Δy values will not vary in the x-direction.

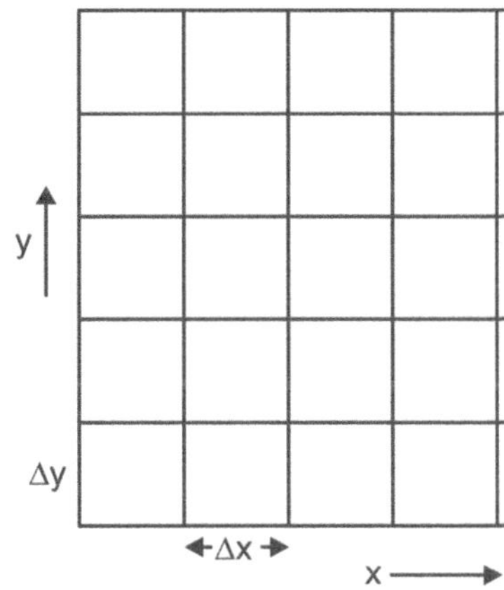

Figure 3.1 Finite difference formulation in the horizontal plane.

Mathematically, $(\Delta x)_{\phi} = (\Delta x)_{equator} \cos \phi$, where ϕ is latitude.

If the system of equations is to be solved for the entire globe, then the equations should be written in spherical coordinate system. However Cartesian coordinate system may be used if the domain dimensions are such that the variations of Δx in y-direction could be neglected. This implies that the domain region is important for the presentation of the various terms.

Here we will consider the grid point system for the Cartesian coordinates for simpler understanding of the numerical methods. The weather prediction system of equations has several partial derivatives, and in the grid point method they are approximated by the differences between finite intervals. As such this procedure is referred to as "finite difference method". In this way, the approximations will transform the partial differential equations into algebraic form that facilitate the solution.

Consider a function $u = u(x)$, for one independent variable, which is the solution of a differential equation. If the bounded region is of length L, it may be divided into a number of segments of equal length Δx. This Δx is called grid distance or grid interval. If the bounded region has J number of grid intervals, and locating the origin at the left end of the x-axis

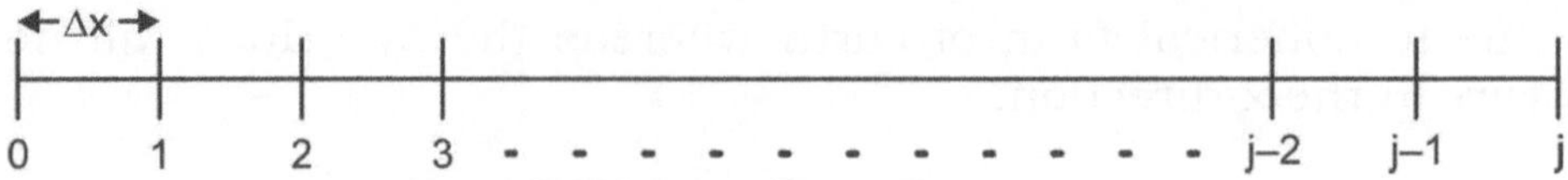

Figure 3.2 Finite difference representation in the x-direction.

The discrete values of u_j describe only a part of the continuous function u(x) due to discretisation. The values at the discrete points j = 0, 1, 2, ….., J, with Δx as the grid interval.

$u_j = u\,(j\,\Delta x)$ where j is a discrete point in x direction

Now the difference of u between two points can be written in different ways for e.g.,

$$\Delta u_j = u_{j+1} - u_j$$
$$\Delta u_j = u_j - u_{j-1}$$
$$\Delta u_j = u_{j+1} - u_{j-1}$$

Writing down the expansion for the u(x) in terms of Fourier series

$$u(x) = \frac{a_0}{2} + \sum_{n>1} (a_n \cos 2\pi\, n\frac{x}{L} + b_n \sin 2\pi\, n\frac{x}{L})$$

The available J+1 values do not permit calculation of all the coefficients a_n, b_n as only J+1 coefficients can be calculated.

That is, we can obtain coefficients for n = 1, 2, 3, …., J/2 ; of which the shortest wavelength will be n = J/2

(i.e) $$\frac{L}{n} = \frac{2L}{J} = \frac{2L}{\frac{L}{\Delta x}} = 2\Delta x$$

This shows that with a grid interval of Δx, it is not possible to resolve waves shorter than with wavelength 2 Δx

We also define wave number, (i.e.) number of waves in the total length L (= J Δx), as

$$K_n = \frac{2\pi n \, \Delta x}{L}$$

where n= 1, 2, 3, ..., J/2; k = 2π/ λ ; λ is wavelength.

3.1 Finite Differences

The choice of taking the difference Δx in relation to the location of the grid points determines the method of finite difference.

For e.g., $\Delta u_j = u_{j+1} - u_j$ forward finite difference

$\Delta u_{j+1/2} = u_{j+1} - u_j$ centered finite difference

In the forward method, difference is taken between the next point (forward direction) and the point at which Δu is calculated whereas in the centered method, difference between the two points on either side is taken.

A differential form of equation can be approximated by replacing with the corresponding finite difference quotients.

For e.g., $$\left(\frac{\partial u}{\partial x}\right)_j = \frac{u_{i+1} - u_i}{x_{i+1} - x_i} = \frac{u_{i+1} - u_i}{\Delta x}$$

It is important to know the consistency of the approximation by examining whether the approximation approaches the derivative as $\Delta x \rightarrow 0$

Using Taylor series expansion, we can write

$$\frac{u_{i+1} + u_i}{\Delta x} = \left(\frac{du}{dx}\right)_j + \frac{1}{2!}\left(\frac{du}{dx}\right)^2 \Delta x + \frac{1}{3!}\left(\frac{du}{dx}\right)^3 (\Delta x)^2 + \ldots$$

If the difference between this and the approximation are examined, it can be seen that the terms: $\frac{1}{2!}\left(\frac{du}{dx}\right)^2 \Delta x$ and higher order are not considered

$$\in = \frac{1}{2!}\left(\frac{du}{dx}\right)^2 \Delta x + \frac{1}{3!}\left(\frac{du}{dx}\right)^3 (\Delta x)^2 + \ldots$$

$\in$ is called truncation error, which indicates the degree of accuracy of the adopted approximation. In this case, $\in = 0(\Delta x)$ denote first order accuracy .

For time analysis, the variation of a variable u with time t is expressed as

$$\frac{du}{dt} \sim \frac{U}{T}$$

where T denotes the time scale and U denote the variability

Further

$$\frac{d^2u}{dt^2} = \frac{d}{dt}\left(\frac{du}{dt}\right) = \frac{U/T}{T} = \frac{U}{T^2}$$

It is important to know about time discretization, since these are applied to time-evolving atmospheric equations

Discretizing with a time step of Δt

$$t^n = t^0 + n\Delta t, n = 1, 2, 3, \ldots.$$

where the superscript indicates the discrete time point

u^n means the value of u at $t = n\Delta t$

We write $\dfrac{du}{dt} = \lim_{\Delta t \to 0} \dfrac{u(t+\Delta t) - u(t)}{\Delta t}$

$$\left|\frac{du^n}{dt}\right| = \frac{u^{n+1} - u^n}{\Delta t}$$

Expanding, using the Taylor series gives

$$u(t+\Delta t) = u(t) + \Delta t\left|\frac{du}{dt}\right| + \frac{1}{2!}\frac{d^2u}{dt^2}(\Delta t)^2 + \frac{1}{3!}\frac{d^3u}{dt^3}(\Delta t)^3 + \ldots$$

which can be written as

$$\frac{du}{dt} = \frac{u(t+\Delta t) - u(t)}{\Delta t} + 0(\frac{\Delta t}{T}\frac{U}{T})$$

The relative error (as divided by U/T), is of the order $\dfrac{\Delta t}{T}$. In order for the finite difference approximation to be acceptable, $\Delta t \ll T$.

As of the previous description, Δx or Δt should be as small as possible for the finite difference approximation to be valid. As we tend to increase the resolution, it is also important to consider higher order approximations and check if they permit the coarser resolution.

Writing

$$u^{n+1} = u^n + \Delta t\left|\frac{du}{dt}\right| + \frac{1}{2!}\frac{d^2u}{dt^2}(\Delta t)^2 + \frac{1}{3!}\frac{d^3u}{dt^3}(\Delta t)^3 + \ldots$$

$$u^{n-1} = u^n - \Delta t\left|\frac{du}{dt}\right| + \frac{1}{2!}\frac{d^2u}{dt^2}(\Delta t)^2 + \frac{1}{3!}\frac{d^3u}{dt^3}(\Delta t)^3 + \ldots$$

Using the two equations, we can derive

$$\left(\frac{du}{dt}\right)^n = \frac{u^n - u^{n-1}}{\Delta t} + 0(\Delta t)$$

$$\left(\frac{du}{dt}\right)^n = \frac{u^{n+1} - u^{n-1}}{2\Delta t} + 0(\Delta t^2)$$

To check whether the 2nd order approximation will help, we consider sinusoidal function as

$$u = \bar{u}\sin 2\pi\frac{t}{T} = \bar{u}\sin\omega t \quad (where\ \omega = \frac{2\pi}{T})$$

Since we know the exact derivative as $\bar{u}\,\omega\cos\omega t$, the errors due to finite difference approximation can be calculated and assessed. We find that both forward and backward methods converge, with decrease of errors as Δt decreases and the 2nd order approximation shows higher order convergence. The calculations also indicate that $\Delta t \le T/8$ which means that at least 8 time points are required to resolve a time scale variation of T so as to keep errors within 10%.

3.2 Finite Difference Schemes

It is important to understand the characteristics of a finite difference scheme with the attributes of consistency, convergence and stability.

We shall consider a simple linear advection equation in the form

$$\frac{\partial u}{\partial t} + c\frac{\partial u}{\partial x} = 0$$

where u = u (x, t) ; c = constant

This equation describes the advection of 'u' at a velocity of c (in the direction of x-axis). For this equation, an analytical solution can be obtained. It can be seen easily that the general solution is

$$u = f(x - c\,t)$$

To verify, at the initial time

$$u(x, 0) = F(x)$$
$$u = F(x - c\,t)$$

Physically, in the x-t plane, the true solution takes constant values along straight line, given by $c = \frac{dx}{dt}$

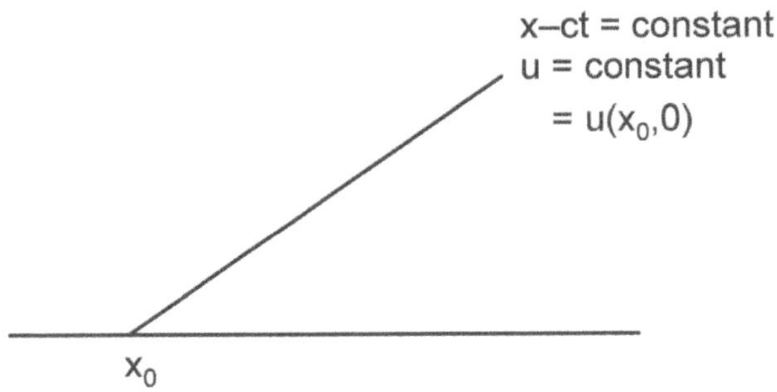

Figure 3.3 Characteristics of the linear advection equation.

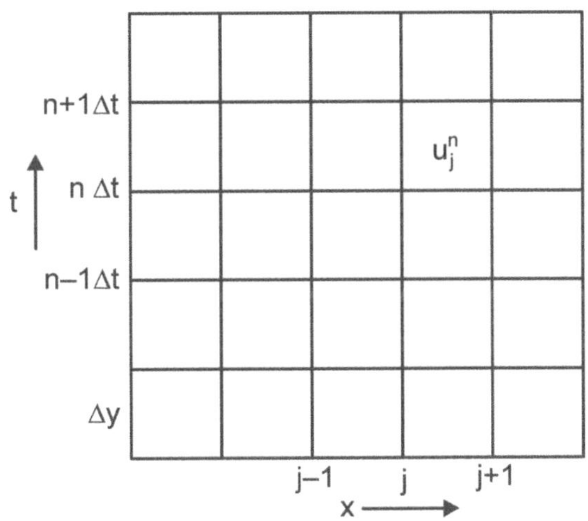

Figure 3.4 Formulation of finite difference grid for the solution of linear advection equation.

Now considering the finite difference solution, using forward difference method, the equation is written as

$$\frac{u_j^{n+1} - u_j^n}{\Delta t} + c\frac{u_j^n - u_{j-1}^n}{\Delta t} = 0$$

where the superscript 'n' denotes time point and subscript 'j' denotes space point.

This scheme is called "forward" or "upstream" scheme meaning the spatial difference is taken as the value at point "j" minus the value at "j-1" where the flow is from "j-1" to "j" which is upstream.

Now the difference between the true and finite difference solution can be written as

$$u_j^n - u(j\Delta x, n\Delta t)$$

The finite difference form can be written as

$$\frac{u(j\Delta x,(n+1)\Delta t) - u(j\Delta x, n\Delta t)}{\Delta t} + c\frac{u(j\Delta x, n\Delta t) - u((j-1)\Delta x, n\Delta t)}{\Delta x} = \in$$

where $\in$ is called the truncation error

Using the Taylor series expansion, we can expand the above equation as

$$\in = \left[\frac{1}{2}\left(\frac{d^2u}{dt^2}\right)\Delta t + \frac{1}{6}\left(\frac{d^3u}{dt^3}\right)\Delta t^2 + \ldots\right] - c\left[\frac{1}{2}\left(\frac{d^2u}{dx^2}\right)\Delta x + \frac{1}{6}\left(\frac{d^3u}{dx^3}\right)\Delta x^2 + \ldots\right]$$

The order of accuracy can be written as

$\in = 0\,(\Delta t) + 0\,(\Delta x)$ or $\in = 0\,(\Delta t, \Delta x)$

This scheme is consistent at first order accuracy.

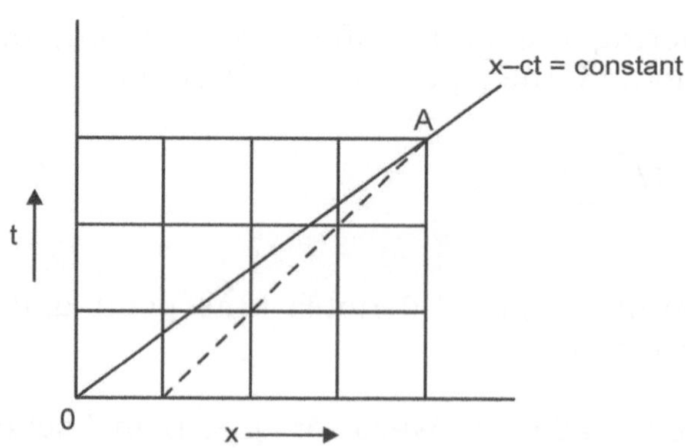

Figure 3.5 Positions of a characteristic and a domain of dependence.

Convergence: The characteristics of a finite difference scheme in terms of convergence are to be ascertained from the behavior of the numerical solutions, when Δx and Δt are made smaller keeping $n\Delta t$ as fixed and for fixed values of Δx and Δt with increase of n (number of time steps). If the error approaches zero as Δx and Δt tend to zero, then the solution is convergent.

From this figure, we note that the characteristic passing through the origin also passes through A, meaning the value of u at A is the same as at O. However, the numerical solution indicates that the value of u at A is different, away from O, as controlled by the values of Δx and Δt. In this case, the errors could be large. It is also to be seen that even if Δt and Δx are reduced proportionately, the same solution would result.

The convergence would be possible only when $c\,\Delta t \leq \Delta x$ (i.e.) the slope of the characteristic is higher than the line OA.

Stability: A numerical solution is considered stable, if the error remains bounded as n increases for fixed values of Δx and Δt. There are three methods, namely "Direct method", "Energy method" and "Von Neumann's method".

In the next part, we will consider the properties of "consistency", "convergence" and "stability" for different time and space finite difference schemes.

3.3 Time Differencing Schemes

The finite difference schemes that are used to approximate the time derivatives are of the second order or sometimes first order only. This is because of the fact that higher order schemes are not as successful with partial differential equations as with ordinary differential equations.

In the partial differential equations, the errors in the numerical solution may be due to inadequacy of the scheme and errors in the initial conditions in contrast to the ordinary differential equations wherein the errors are due to the applied scheme only. Another reason is that, for ascertaining the stability, a smaller time step is required.

We consider the equation

$$\frac{du}{dt} = f(u,t); u = u(t) \qquad \text{[variable t is time]}$$

Let us denote u^n is the value of u at $n\Delta t$,

Assuming that the value of u^1, u^2, u^3, - - -, u^{n-1}, u^n are known, a time differencing scheme to obtain the value u^{n+1} is to be designed.

3.3.1 Two-level Schemes

These are the schemes, in which the values of the variable at two time steps n and n+1 are only related. For this reason, at the first time step only these schemes can be used. The mathematical expression is written as

$$u^{n+1} = u^n + \int_{n\Delta t}^{(n+1)\Delta t} f(u,t)\,\Delta t$$

The finite difference form following the Euler scheme (Forward Scheme) is

$$u^{n+1} = u^n + f^n \Delta t \quad \text{where } f^n = f(u^n, n\,\Delta t)$$

The truncation error is of 1st order, O (Δt). In this scheme, f is taken at n Δt and not centered. In general, uncentered schemes are 1st order accurate and centered schemes are 2nd order accurate.

3.3.2 Backward Scheme

The equation for this scheme is expressed as

$$u^{n+1} = u^n + f^{n+1}\,\Delta t$$

In this scheme, f corresponds to n+1 Δt as of u^{n+1} and so the scheme is called implicit. Solution of this requires solving of a set of simultaneous equations, with one equation corresponding to one grid point.The truncation error is of the order, O (Δt)

3.3.3 Trapezoidal Scheme

$$u^{n+1} = u^n + \frac{1}{2}\Delta t\,(f^n + f^{n+1})$$

In this scheme, the values of f at the beginning and end of integration are averaged. This is also an implicit scheme due to the presence of f^{n+1}. The truncation error is O (Δt^2).

3.3.4 Matsuno Scheme (Euler backward scheme)

This scheme uses three steps.

First the value of u^{n+1*} is obtained using

$$u^{n+1*} = u^n + f^n \Delta t$$

Then $u^{\overline{n+1}*}$ is used to compute an approximation of f^{n+1} as

$$f^{\overline{n+1}*} = f[u^{\overline{n+1}*}, (n+1)\Delta t]$$

Finally $$u^{n+1} = u^n + f^{\overline{n+1}*}.\Delta t$$

This is an explicit scheme, and first order accurate.

3.3.5 Heun Scheme

$$u^{\overline{n+1}*} = u^n + f^n \Delta t$$
$$f^{\overline{n+1}*} = f[u^{\overline{n+1}*}, (n+1)\Delta t]$$
$$u^{n+1} = u^n + \frac{1}{2}\Delta t\left[f^n + f^{\overline{n+1}}*\right]$$

This scheme is an explicit scheme, similar to trapezoidal scheme.

The above five schemes are two level schemes. Now two 3-level schemes will be presented.

3.4 Three-level Schemes

3.4.1 Leapfrog Scheme

$$u^{n+1} = u^{n-1} + \int_{n-1}^{n+1} f(u,t)\Delta t$$

In this scheme, centered integration is done, from one time step back to one time step forward.

The truncation error is of 2nd order O (Δt^2). This is the most widely used scheme in weather prediction models.

3.4.2 Adams - Bashforth Scheme

$$u^{n+1} = u^n + [\frac{3}{2} f^n - \frac{1}{2} f^{n-1}]\Delta t$$
$$= u^n + [f^n + \frac{f^n - f^{n-1}}{2}]\Delta t$$

In this scheme, f^n is estimated at the central point Δt, by extrapolating the difference between f^n and f^{n-1}. This is also second order accurate scheme.
Now we consider finite difference schemes for space differentials.

3.5 Schemes for Advection Equation

Consider the 2-dimesnional linear advection equation

$$\frac{\partial u}{\partial t} + c\frac{\partial u}{\partial x} = 0 \; ; \quad c = \text{constant}$$
$$u = u\ (x, t)$$

The general solution is u = f (x – c t)

$$\frac{\partial u_j}{\partial t} = -c\frac{u_{j+1} - u_{j-1}}{2\Delta x}$$

3.5.1 Leapfrog Scheme

By considering Leapfrog scheme for time differencing

$$\frac{u_j^{n+1} - u_j^{n-1}}{2\Delta t} = -c\frac{u_{j+1}^n - u_{j-1}^n}{2\Delta x}$$

Representing u_j in terms of harmonic components

$$u_j = u(t)e^{ikj\Delta x}$$

From this we get $\dfrac{du}{dt} = i\left[-\dfrac{c}{\Delta x}\sin k\Delta x\right]u$

Let $\quad \omega = -\dfrac{c}{\Delta x}\sin k\Delta x$

$$u^{n+1} = u^{n-1} + 2i\left(-c\frac{\Delta t}{\Delta x}\sin k\Delta x\right)u^n$$

For stability $\left|-\dfrac{c\Delta t}{\Delta x}\sin k\Delta x\right| \leq 1$

Since $|\sin k\Delta x| \leq 1$ the criterion reduces to $c\dfrac{\Delta t}{\Delta x} \leq 1$

This criterion, already discussed earlier, implies that reduction of Δt and Δx may not lead to stability, but the ratio of $\dfrac{\Delta t}{\Delta x}$ should be reduced enough to ensure stability. This condition was found by Courant-Friedrichs-Lewy in 1928 and is referred to as Courant-Friedrichs-Lewy (CFL) stability criterion.

3.5.2 Matsuno Scheme

In this scheme, as a first step, approximate values of u_j^{n+1*} are calculated using forward scheme.

$$\frac{u_j^{n+1*} - u_j^n}{\Delta t} = -c\frac{u_{j+1}^n - u_{j-1}^n}{2\Delta x}$$

Second step is to use them as in backward scheme

$$\frac{u_j^{n+1} - u_j^n}{\Delta t} = -c\frac{\overline{u_{j+1}^{n+1*}} - \overline{u_{j-1}^{n+1*}}}{2\Delta x}$$

The * terms on the right hand side can be eliminated using the earlier equation.

$$\frac{u_j^{n+1}-u_j^n}{\Delta t}=-\frac{c}{2\Delta x}\left(u_{j+1}^n-\frac{c\Delta t}{2\Delta x}\left(u_{j+2}^n-u_j^n\right)\right)$$
$$+\frac{c}{2\Delta x}\left[u_{j-1}^n-\frac{c\Delta t}{2\Delta x}\left(u_j^n-u_{j-2}^n\right)\right]$$
$$=-\frac{c}{2\Delta x}\left(u_{j+1}^n-u_{j-1}^n\right)$$
$$+\frac{c^2\Delta t}{(2\Delta x)^2}\left(u_{j+2}^n-2u_j^n+u_{j-2}^n\right)$$

Without the second term on the right hand side, the finite difference equation is same as of the forward scheme. This second term on right hand side $\rightarrow 0$ as Δt and $\Delta x \rightarrow 0$ and so is a consistent scheme.

3.6. Energy Method and Stability

The energy method is useful for wide applications, and more so for nonlinear equations.
In this method, if we know that the true solution is bounded, we test if $\Sigma\left(u_j^n\right)^2$ is also bounded. If it is bounded, then every value of u_j^n is bounded and stability of the scheme is established.

Many of the numerical schemes are of the form

$$u_j^{n+1}-u_j^n=-\frac{1}{2}\mu\left(\overset{*}{u}_{j+1}-\overset{*}{u}_{j-1}\right) \qquad .....(3.1)$$

where $\mu = \frac{c\,\Delta t}{\Delta x}$ and u_j^* is a linear function of a number of values u_j^n.

Two level non-iterative schemes are of the form

$$u_j^* = \alpha u_j^n + \beta u_j^{n+1}$$

Iterative two level schemes have the form

$$u_j^* = u_j^n - \frac{\beta}{2}\mu\left(u_{j+1}^n - u_{j-1}^n\right)$$

For Adams-Bashforth scheme

$$u_j^* = \frac{3}{2}u_j^n - \frac{1}{2}u_j^{n-1}$$

To study the stability, we multiply equation (1) with u_j and sum for all j, we will have

$$\sum_j u_j^*\left(u_j^{n+1} - u_j^n\right) = \frac{1}{2}\mu\sum_j u_j^*\left(u_{j+1}^* - u_{j-1}^*\right)$$

The term on right hand side is vanished, if cyclic boundary conditions are assumed. Then

$$\sum_j u_j^*\left(u_j^{n+1} - u_j^n\right) = 0$$

This relation leads to

$$\sum_j \frac{1}{2}\left[\left(u_j^{n+1}\right)^2 - \left(u_j^n\right)^2\right] = \sum_j \frac{1}{2}\left[\left(u_j^{n+1} - u_j^n\right)\left(u_j^{n+1} + u_j^n\right)\right]$$

Using the relation $u_j^* = \alpha u_j^n + \beta u_j^{n+1}$

and $\beta = (1-\alpha)$

We get

$$\sum_j \frac{1}{2}\left[\left(u_j^{n+1}\right)^2 - \left(u_j^n\right)^2\right] = (\alpha - \frac{1}{2}) \sum_j \left[\left(u_j^{n+1} - u_j^n\right)^2\right]$$

If $\alpha > \frac{1}{2}$ Unstable; $\alpha = \frac{1}{2}$ stable and neutral ; $\alpha < \frac{1}{2}$ Stable and damping.

3.7 Lax-Wendroff Scheme

This scheme is completely different from the other schemes.

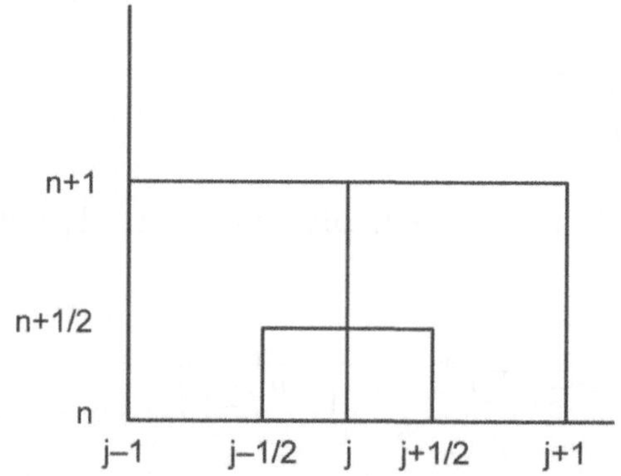

Figure 3.6 Space-time finite difference formulation of the Lax-Wendroff scheme.

In this scheme, centered space and forward time differencing is used, and taking arithmetic averages of the two points on either side.

Step wise computation is as follows

$$u_{j+\frac{1}{2}}^{n+\frac{1}{2}} = \frac{1}{2}\left[u_{j+1}^n + u_j^n\right] - \frac{\Delta t}{2\Delta x}\left[u_{j+1}^n - u_j^n\right]$$

The first term on the right hand side is the average at j and j+1, corresponding to $j+\frac{1}{2}$ *i.e.* $u^{n}_{j+\frac{1}{2}}$

$$\frac{u^{n+\frac{1}{2}}_{j+\frac{1}{2}} - u^{n}_{j+\frac{1}{2}}}{\Delta t}$$ is the time difference (forward) at $u^{n}_{j+\frac{1}{2}}$

$$\frac{u^{n}_{j+1} - u^{n}_{j}}{2\Delta x}$$ correspond to centered space difference at $u^{n}_{j+\frac{1}{2}}$

Second step: Similarly

$$u^{n+\frac{1}{2}}_{j-\frac{1}{2}} = \frac{1}{2}\left[u^{n}_{j} + u^{n}_{j-1}\right] - \frac{\Delta t}{2\Delta x}\left[u^{n}_{j} + u^{n}_{j-1}\right]$$

3rd Step: Leapfrog half step

$$u^{n+1}_{j} = u^{n}_{j} - \frac{\Delta t}{\Delta x}\left[u^{n+\frac{1}{2}}_{j+\frac{1}{2}} - u^{n+\frac{1}{2}}_{j-\frac{1}{2}}\right]$$

$$\left[\frac{u^{n+1}_{j} - u^{n}_{j}}{\Delta t}\right]$$ correspond to $\frac{\partial u}{\partial t}$ at t = n+1/2 and at x = j

and $$\left(\frac{\partial u}{\partial x}\right)^{n+\frac{1}{2}}_{j} = \frac{u^{n+\frac{1}{2}}_{j+\frac{1}{2}} - u^{n+\frac{1}{2}}_{j-\frac{1}{2}}}{\Delta x}$$

Now substituting for $u_{j+\frac{1}{2}}^{n+\frac{1}{2}}$ and $u_{j-\frac{1}{2}}^{n+\frac{1}{2}}$ from the 1st and 2nd steps, we get

4th Step:

$$u_j^{n+1} = u_j^n - \frac{c\Delta t}{2\Delta x}\left[u_{j+1}^n - u_{j-1}^n\right] + \frac{1}{2}\frac{c^2\Delta t^2}{\Delta x^2}\left[u_{j+1}^n - 2u_j^n + u_{j-1}^n\right] \quad .....(3.2)$$

The last term on right hand side reduces to zero as $\Delta t \to 0$ and $\Delta x \to 0$.

If $\Delta x \to 0$ and Δt is fixed, the term represents damping.

Stability Properties:

$$u_j^n = u\ (n)\ e^{i\ k\ j\Delta x}$$

substituting from previous equation (3.2)

$$u^{n+1} = (1 - \mu^2(1 - \cos k\Delta x) - i\mu \sin k\Delta x)\, u^n$$

Let $\qquad \lambda = 1 - \mu^2(1 - \cos k\Delta x) - i\mu \sin k\Delta x$

Using trigonometric identities

$$\cos k\Delta x = 1 - 2\sin^2 \frac{k\Delta x}{2}$$

$$\sin k\Delta x = 2\sin\frac{k\Delta x}{2}\cos\frac{k\Delta x}{2}$$

We get $|\lambda| = \left[1 - 4\mu^2(1-\mu^2)\sin^4 \frac{k\Delta x}{2}\right]^{1/2}$

Since $\frac{k\Delta x}{2} \leq 1$; so right hand side is always greater than $1 - 4\mu^2(1-\mu^2)$

If $\mu = 1$ $\lambda = 1$;
if $\mu^2 = 1/2$; λ is minimum

For $|\lambda| \leq 1$ the scheme is stable if $\frac{c\Delta t}{\Delta x} \leq 1$ (CFL criterion) and for a given wavelength, the amplification factor is given by

$$\lambda_{\min} = \left(1 - \sin 4k\Delta x/_2\right)^{1/2}$$

If $k\,\Delta x = \pi/2$; $\lambda = 0$

$\lambda \rightarrow 1$ as $k \rightarrow 0$ (i.e.) as the wavelength $\rightarrow \infty$

So the amount of damping is longer for short waves.

Lax-Wendroff scheme is generally used in atmospheric models as it is second order accurate, explicit, not unconditionally unstable and has no computational mode.

3.8 Dispersion

The linear advection equation,

$$\frac{\partial u}{\partial t} + c\frac{\partial u}{\partial x} = 0 \qquad .....(3.3)$$

has no dispersion, as all wave modes have the same phase speed c.

Considering

$$\frac{\partial u_j}{\partial t}+c\frac{u_{j+1}-u_{j-1}}{2\Delta x}=0 \qquad .....(3.4)$$

in which space differencing is done using centered finite difference method.

$$u_j = u(t)e^{ikj\Delta x}$$

provided that $\frac{du}{dt}+ik\left(c\frac{\sin k\Delta x}{k\Delta x}\right)u=0$

substitution into equation (3.3) gives

$$\frac{du}{dt}+ikcu=0$$

The solutions of equation (3.4) propagate into the phase speed c^*

$$c^* = c\frac{\sin k\Delta x}{k\Delta x}$$

where k is wave number, meaning c^* is a function of wave number. So the finite differencing leads to dispersion of waves.

As k Δx increases from zero, c^* decreases as compared to c and becomes zero when $k\Delta x = \pi$ (i.e.) for the shortest wavelength of 2 Δx.

This expression also indicates that all waves propagate at lesser speeds than the true speed, the deceleration increasing as wavelength decreases.

3.9 Group Velocity

Group velocity is defined as

$$c_g = \frac{d\omega}{dk} = \frac{d\,kc}{dk}$$

where $\omega = \omega(k)$ is the frequency and k is the wave number.

$$c_g^* = \frac{d\,kc^*}{dk} = c\cos k\Delta x$$

Since, in the linear advection, c is a constant and so $c_g = c$

Since $$c^* = c\frac{\sin k\Delta x}{k\Delta x}$$

This shows that as k Δx increases from zero, the group velocity decreases from c_g and become equal to $-c_g$ for the shortest wavelength of 2 Δx.

Part B. Aliasing and nonlinear instability

Consider a simple differential equation with a nonlinear term (as similar to in weather prediction system)

$$\frac{\partial u}{\partial t} + u\frac{\partial u}{\partial x} = 0$$

For which u = f (x – u t) is the general solution. When multiplication of terms is considered in finite difference form, certain errors arise.

Consider u = sin k x

With grid interval of Δx, 2Δx is the shortest resolvable wavelength and the wave number

$$K_{max} = \frac{2\pi}{2\Delta x} = \frac{\pi}{\Delta x}$$

Considering u = sin k x, where $k < k_{max}$

$$u\frac{\partial u}{\partial x} = \sin kx\; k \cos kx = k \sin kx \cos kx$$
$$= \frac{1}{2} k \sin 2\,k\,x$$

From this we analyse that to the wave number $1/2k_{max} < k \leq k_{max}$ the nonlinear term will generate a wave number that cannot be resolved.

We recognize that a wave with $k > k_{max}$ will be seen as a wave with $k < k_{max}$

From trigonometry,

$$\sin k\, x = \sin\,[2k_{max} - (2k_{max} - k)]\, x$$

Substituting $k_{max} = \dfrac{\pi}{\Delta x}$ and taking $x = j\,\Delta x$

$$\sin k\, j\,\Delta x = -\sin\,[2k_{max} - k]\, j\,\Delta x$$

We realize that, with the representation of the above grid points, wave numbers $2k_{max} - k$ and k cannot be distinguished. So, for waves with $k > k_{max}$ they will be reflected as

$$k^{*} = 2k_{max} - k$$

For e.g., $L = \frac{4\Delta x}{3}$ meaning $k = \frac{3\pi}{2\Delta x}$ which is higher than $k_{max} = \frac{\pi}{\Delta x}$; the same will now be seen as a wave with wave number $[\dfrac{2\pi}{\Delta x} - \dfrac{3\pi}{2\Delta x}] = \dfrac{\pi}{2\Delta x}$ (i.e.) with a wavelength $4\Delta x$ (Figure 3.7).

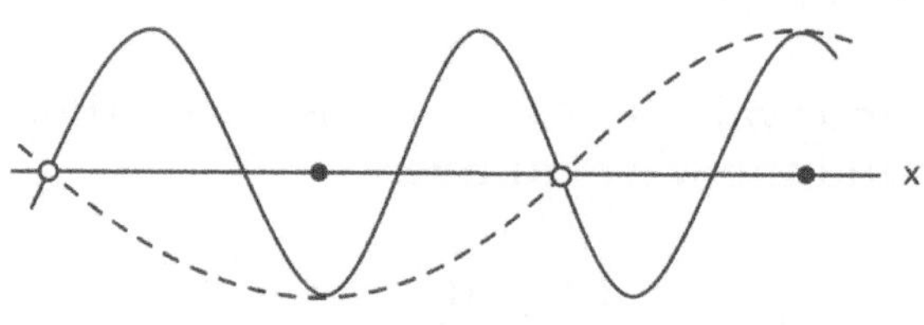

Figure 3.7 False representation of a wave with wavelength of $4\Delta x/3$ as a wave of wavelength $4\Delta x$.

The above described errors that arise due to the application of finite difference approximation to estimate nonlinear terms resulting in false representation of waves is called aliasing.

Considering a spectrum of scales of atmospheric motion, we observe that most of the energy is contained in smaller wave numbers of mostly 4 to 7 and the energy tapers off as the wave number goes beyond 10 or so.

Due to use of finite difference approximation method, the shape of the wave spectrum changes due to aliasing errors. Due to nonlinear terms, the energy associated with waves of $k > k_{max}$ will be given to waves with wave numbers near k_{max} (smaller waves) which will grow masking the real waves with higher wavelengths (or smaller wave numbers) leading to very large errors spoiling the entire prediction. This phenomenon arising due to nonlinear terms is called "nonlinear instability".
In numerical weather prediction, if the model integration is to be carried out for a longtime period, nonlinear instability has to be suppressed. One of the suggestions was to remove the 1/3 of the waves near k_{max} (i.e.) removing of $k > 2/3$ as this will effect elimination of aliases of $k>2/3$. Another suggestion was to use Lax-Wendroff finite difference scheme at different intervals. Yet another suggestion is to use a dissipative term.

Arakawa has proposed the best approach based on conservation of energy. Brief mathematics considering the vorticity equation elaborates this method.

$$\frac{\partial\zeta}{\partial t} + V.\nabla\zeta = 0$$

$$\zeta = \nabla^2\psi$$

$V = k \times \nabla\psi$; [assuming wind is non-divergent]

$$\frac{\partial\nabla^2\psi}{\partial t} = J\left(\psi, \nabla^2\psi\right)$$

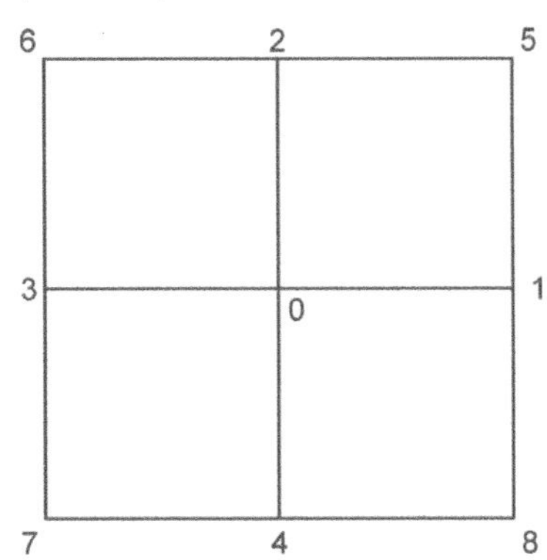

Figure 3.8 Representation of grid numbers for the computation of Jacobian.

The right hand side term denotes the nonlinear term, expressed as Jacobian term with finite difference form. Mathematically, Jacobian term is written as

$$J(A,B) = \frac{\partial A}{\partial x}\frac{\partial B}{\partial y} - \frac{\partial A}{dy}\frac{\partial B}{dx}$$

$$= \frac{\partial}{\partial y}\left[B\frac{\partial A}{\partial x}\right] - \frac{\partial}{\partial x}\left[B\frac{\partial A}{\partial y}\right]$$

$$= \frac{\partial}{\partial x}\left[A\frac{\partial B}{\partial y}\right] - \frac{\partial}{\partial y}\left[A\frac{\partial B}{\partial x}\right]$$

The finite difference forms of the 3 different Jacobians (Figure 3.10) are written as

$$j^{++}(A,B) = \frac{1}{4d^2}\left[(A_1 - A_3)(B_2 - B_4) - (A_2 - A_4)(B_1 - B_3)\right]$$

$$j^{x+}(A,B) = \frac{1}{4d^2}\left[(B_2(A_5 - A_6) - B_4(A_8 - A_7) - B_1(A_5 - A_8) + B_3(A_6 - A_7))\right]$$

$$j^{+x}(A,B) = \frac{1}{4d^2}\left[(A_1(B_5 - B_8) - A_3(B_6 - B_7) - A_2(B_5 - B_6) + A_4(B_8 - B_7))\right]$$

Arakawa demonstrated that

$$J = \frac{1}{3}\left(J^{++} + J^{x+} + J^{+x}\right)$$

J will conserve average vorticity, kinetic energy and enstrophy and so prevents nonlinear instability.

3.10 Staggered Grids

When primitive equation models are used for weather prediction, propagation of gravity waves will also be present. Accordingly, finite difference solutions need to simulate two processes appropriately, one is the geostrophic adjustment process which establishes quasi-nondivergent state due to dispersion of inertia- gravity waves and the other is proper simulation of the quasi-nondivergent state after it is established.

From the mathematical considerations, we know the propagating speed of gravity waves is $\pm\sqrt{gH}$ as derived using the mathematical expressions for u (velocity) and h (depth) variables as

(g, H are constants) $\frac{\partial u}{\partial t} = -g\frac{\partial h}{\partial x}$ and $\frac{\partial h}{\partial t} = -H\frac{\partial u}{\partial x}$

Using finite differences

$$\frac{\partial u_j}{\partial t} = -g\frac{h_{j+1} - h_{j-1}}{2\Delta x};\ \frac{\partial h_j}{\partial t} = -H\frac{u_{j+1} - u_{j-1}}{2\Delta x}$$

In this problem, we have two dependent variables u and h, and their finite differences are calculated at every grid point. Having both the variables identified at the same grid point (as shown in Figure 3.9a), which are completely decoupled, are to be solved. So, it is suggested to carry out the solutions at alternate points (as shown in Figure 3.9b). This is called staggered grid. For this grid, the computational time would be reduced by half and with the same truncation error.

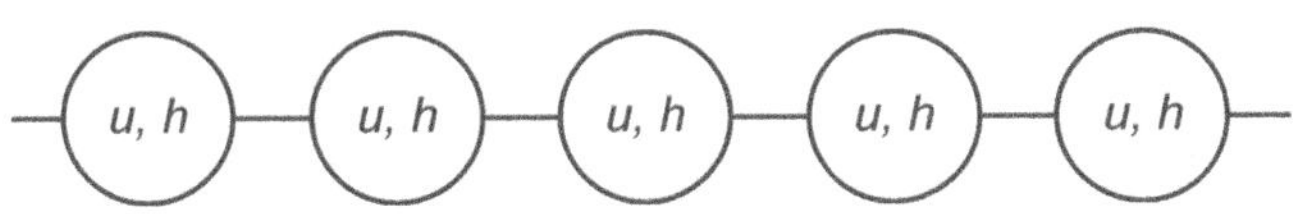

Figure 3.9(a) A grid with two variables at every grid point.

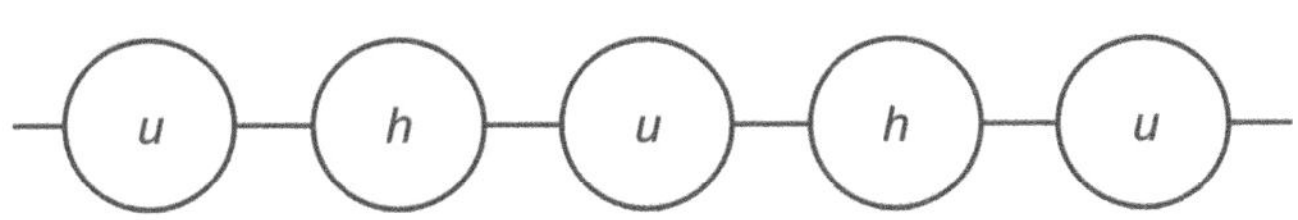

Figure 3.9(b) A grid with two variables at alternate grid points

Analogously, if we consider two dimensional gravity waves, we will have

$$\frac{\partial u}{\partial t} = -g\frac{\partial h}{\partial x};\frac{\partial v}{\partial t} = -g\frac{\partial h}{\partial y};\frac{\partial h}{\partial t} = -H(\frac{\partial u}{\partial x} + \frac{\partial v}{\partial y})$$

Use of primitive equations allow two types of motion, which are low-frquency quasi-geostrophic and quasi- non divergent flow and high-frequency inertia-gravity waves. The inertia-gravity waves are generated continuosly but then energy at small scale will be dispersed and this process is called geostrophic adjustment. The remaining motion is quasi-geostrophic and changes slowly. For correct simulation of the quasi-geostrophic process, five types of finite differencing grids in two dimensional space are suggested. Although the computational time will nearly be the same, the properties of solutions differ. The five grid models are shown in figure 3.10.

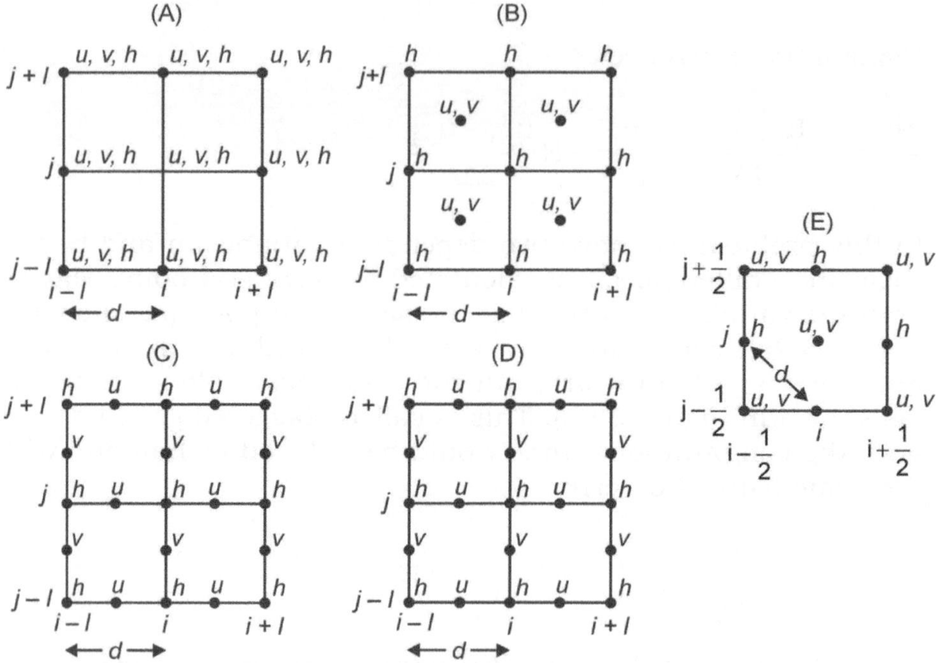

Figure 3.10 Five types of staggered grid representation.

The mathematics yields wave solutions for the five methods, which are given as

$$\left(\frac{\upsilon}{f}\right)^2 = 1 + \left(\frac{\lambda}{d}\right)^2 \sin^2 kd$$

$$\left(\frac{\upsilon}{f}\right)^2 = 1 + 4\left(\frac{\lambda}{d}\right)^2 \sin^2 \frac{kd}{2}$$

$$\left(\frac{\upsilon}{f}\right)^2 = \cos^2 \frac{kd}{2} + 4\left(\frac{\lambda}{d}\right)^2 \sin^2 \frac{kd}{2}$$

$$\left(\frac{\upsilon}{f}\right)^2 = \cos^2 \frac{kd}{2} + \left(\frac{\lambda}{d}\right)^2 \sin^2 kd$$

$$\left(\frac{\upsilon}{f}\right)^2 = 1 + 2\left(\frac{\lambda}{d}\right)^2 \sin^2\frac{kd}{\sqrt{2}}$$

where k is wave number, λ is the Rossby radius of deformation, d is the grid interval and f is Coriolis parameter, υ is nondimensional frequency

The dispersion properties in terms of υ / f and $k d / \pi$ are hown in figure 3.11.

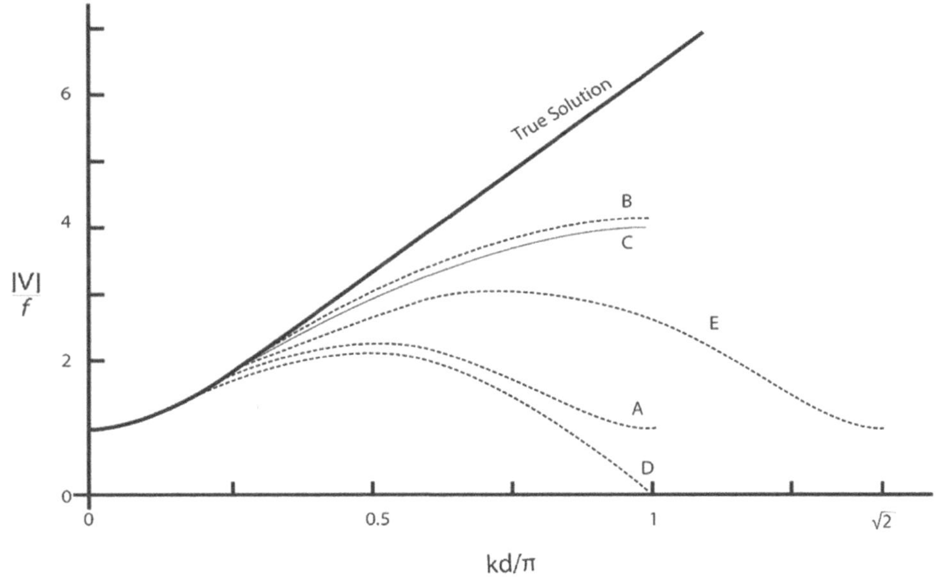

Figure 3.11 Functions of v/f in relation to kd/π for λ/d=2 (Mesinger and Arakawa, 1976)

The figure clearly shows the limitations of A and D grid types. The other three grid types are better. The differences between B and E are good to compare as the grids differ by rotation of 45°. The B and E grid types have a problem of producing false low frequencies of short waves. Arakawa considered that C-grid is the best to simulate geostrophic adjustment process.

Part C. Spectral Method

The spectral method for meteorological applications was first introduced by Silberman (1954). Very few applications of this method had followed because of the computer storage space and computation time required for the calculation of the interaction coefficients of the nonlinear terms in the governing equations. The innovative transform method envisaged by Orszag (1970) and Eliassen et al. (1970), which facilitated easier computation of the nonlinear term, regained the application of spectral method. The first successful applications of spectral atmospheric models were made by Machenhauer and Daley (1972) and by Bourke (1974). Now almost all of the global atmospheric models use spectral method.

In this method, the spatial variations of a variable are expressed as a sum of finite number of waves. The forecast variables are expressed as continuous functions and their horizontal derivatives are also expressed as derivatives of these functions. Because of the wave type representation, information of the selected number of waves is fully retained and will not have errors as of finite difference representation. However, spectral models use calculations of advection (nonlinear) terms and physics parameterizations at grid points. This implies that the model variables are to be transformed backward and forward from spectral to physical and physical to spectral domains.

For the grid point calculations, the spectral coefficients of variables are transformed to grid points. The exact location and spacing of the grid points is dependent on the maximum number of waves represented in spectral space. Since the distribution of the grid points match the spectral resolution, the nonlinear terms are calculated precisely. However, since model physics are also calculated on the same grid, errors in the calculation of local physics and their transformation between the physical-spectral-physical representations may induce errors.

Since a part of the calculations are performed on the model grid, errors may be induced in the calculation of gradients; arising of errors resulting due to distortions from physical processes (which may not have wave like structure and sometime step-like such as latent heat release at certain locations) in time tendencies of forecast variables; and differences in patterns of variables due to physical processes at adjacent grid points leading to oscillations of short wavelengths. To alleviate this problem, spectral models use some filtering methods.

In this method, the atmospheric fields (on a sphere) are expressed in terms of series expansion of functions, called spherical harmonics. These spherical harmonics are the eigenfunctions of the Laplace operator on the sphere, which are complex functions of longitude λ (varying from 0 to 2π) and μ (=sin ϕ, ϕ is latitude, varying from -1 to +1) and are expressed by the mathematical relation as

$$\nabla^2 Y_{m,n}(\lambda, \mu) = -\frac{n(n+1)}{a^2} Y_{m,n}(\lambda, \mu)$$

where a is radius of earth; and $Y_{m,n}$ are the spherical harmonics.

When normalized, these functions form orthonormal basis and verify the relation as

$$\frac{1}{4\pi} \int_{-1}^{+1} \int_{0}^{2\pi} Y_n^m(\lambda, \mu) Y_n^{m*}(\lambda, \mu)\, d\lambda\; d\mu = \delta$$

where Kronecker delta δ=0 or 1 depending on n, m and n′ and m′ (here m′ and n′ are the coefficients of Y_n^{m*})

δ=0 when n≠n′ and $m \neq m'$
δ=1 when n=n′ and m=m′

The spherical harmonics are expressed as the product of two functions, one of the variable λ and second of the variable μ.

$$Y_n^m(\lambda, \mu) = P_n^m(\mu) e^{im\lambda}$$

P_n^m are known as Associate Legendre functions of the first kind with degree 'n' and order 'm'.

On a circle of latitude, Y_n^m is periodic and so 'm' can be interpreted as zonal wave number. Interpretation of 'n' is more difficult as 'n-m' represent the number of nodes in meridional direction. Figure 3.12 depicts the representation of m and n. In the figure, maximum of m is taken as 5. For m=0, n-m will be 5 and there will be 5 nodes in the meridional direction and no flow will be zonal in the zonal direction; for m=1, n-m=4 which means one wave in zonal direction and 4 nodes in the

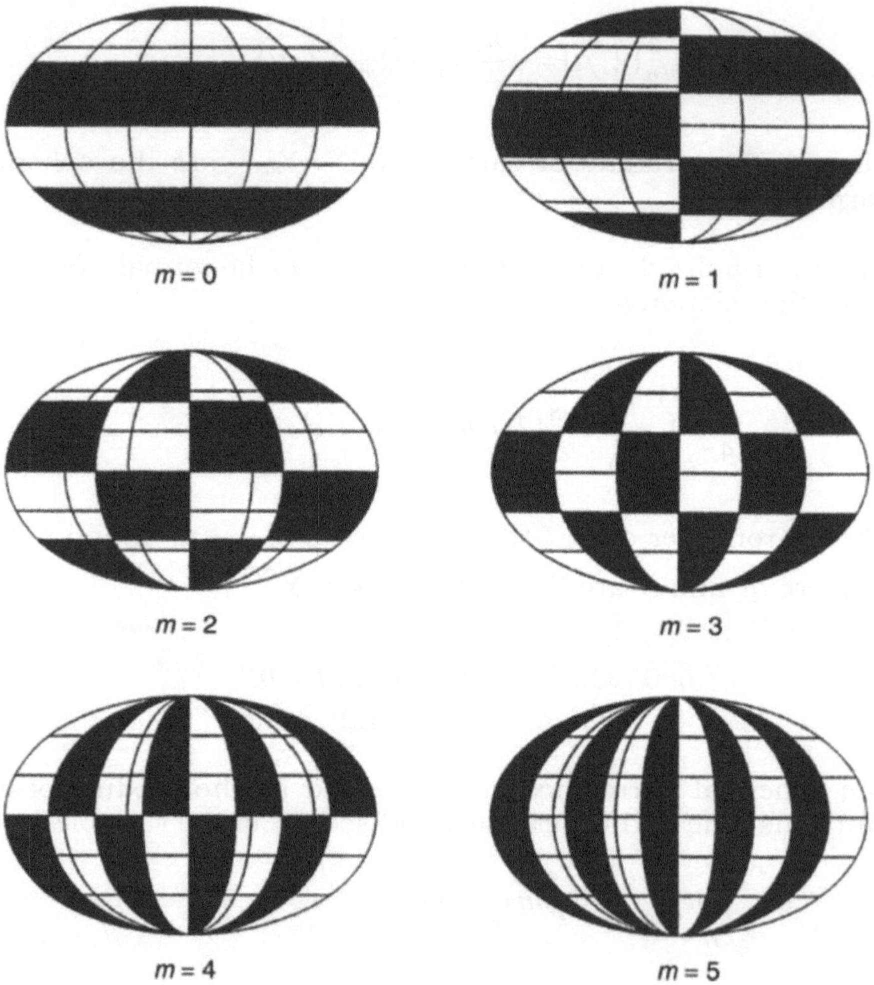

Figure 3.12 Schematic of the real part of the spherical harmonics (positive values in black and negative values in white) for n=5. The harmonics corresponding to m=0 (top left) are zonal, those corresponding to n=m (bottom right) are sectorial, and the others are tesseral harmonics. (Baer, 1972, J. Atmos. Sci., 29, 651, Amer. Met. Soc.)

meridional direction, ..., for m=5, n-m=0 which means five waves in zonal direction and zero nodes in meridional implying no variation in y-direction. The representation of 'm' can be clearly seen for its different values from 0 to 5. Similarly 'n-m' shows the number of nodes in the meridional direction.

$P_n^m(\mu)$ are known as Associate Legendre functions of first kind of degree 'n' and order 'm'. P_n^m (μ) are the real functions of μ, and the solutions of the Legendre equation.

$$\frac{d}{d\mu}\left[\left(1-\mu^2\right)\frac{d}{d\mu}P_n^m(\mu)\right]+\left[n(n+1)-\frac{m^2}{\left(1-\mu^2\right)}\right]P_n^m(\mu)=0$$

Rodrigues formula gives the expression for P_n^m as

$$P_n^m(\mu)=\frac{\left(1-\mu^2\right)^{m/2}}{2^n n!}\frac{d^{n+m}}{d\mu^{n+m}}\left(\mu^2-1\right)^n \qquad .....(3.5)$$

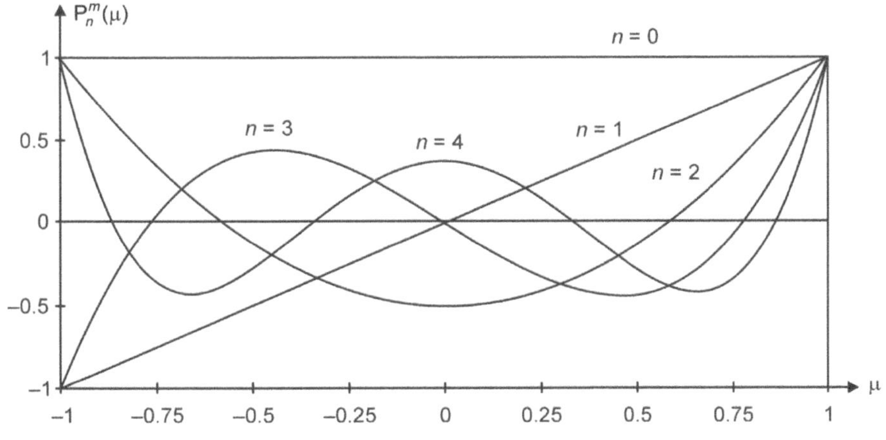

Figure 3.13 shows the variation of P_n^m (μ) as a function of μ, for typical values of m=0, and for n=0, 1, 2, 3 and 4. The magnitude of n-$|m|$ corresponds to the number by zeros of the function $P_n^m(\mu)$ in the interval -1 to +1 and sometimes interpreted as nodes.

As seen from the figure 3.13, P_n^m (μ) takes the values from -1 to +1, for n=0 all the values of P_n^m (μ) are equal to +1; for n=1, P_n^m (μ) varies from -1 to +1 linearly; for n=2, P_n^m (μ) varies from +1 passes through to –ve values and again to +ve values passing through zero values twice to end up at +1; for n=3, P_n^m (μ) values vary from -1 passing through to +ve values and then decline towards –ve and again increases to +ve finally attaining +1; and for n=4, P_n^m (μ) values start at +1 and passes through –ve and +ve regimes alternately to end at +1. These indicate that the value for n would indicate number of zeros (or nodes) (i.e.) the passage of alternating regimes. Odd and even numbers for n indicate their origin at -1 or +1 respectively although all of them have end value at +1. This supplements the previous information of the representation of number of nodes in the meridional direction.

Further, considering the properties of the Associate Legendre function P_n^m (μ), we see that the orthogonality relation leads to

$$\frac{1}{2}\int_{-1}^{+1} P_n^m P_{n'}^m \, d\mu = 0 \text{ for } n \neq n'$$

From equation (1), it can be seen that P_n^m (μ) is defined only for $n \geq m$ and $m \geq 0$

So $P_n^m(\mu)=0$ for $n \leq m$

and $P_n^{-m}(\mu) = P_n^m(\mu)$ for positive values of m

This gives us

$$Y_n^{-m}(\lambda,\mu) = Y_n^{m*}(\lambda,\mu)$$

where superscript * indicates the complex conjugate

$Y_n^m(\lambda,\mu)$ can be subdivided as even and odd harmonics characterized by n-$|m|$.

Even harmonics are symmetric with respect to equator whereas odd harmonics are antisymmetric.

$$Y_n^m(\mu) = -Y_n^m(\mu)$$

When a field is designed to be symmetric, they can be expanded using only the even harmonics; and similarly antisymmetric fields are expanded using only the odd harmonics.

In atmospheric modeling, geopotential , temperature, pressure, zonal wind and velocity potential are taken as symmetric, whereas meridional wind and stream function are considered as antisymmetric.

As part of orthonormality relation, P_n^m (μ) satisfy

$$\frac{1}{2}\int_{-1}^{+1} P_n^m P_n^m d\mu = 1$$

So the normalized Legendre function is written as

$$P_n^m(\mu) = \sqrt{(2n+1)}.\sqrt{\frac{(n-m)!}{(n+m)!}}\frac{\left(1-\mu^2\right)^{m/2}}{2^n n!}\frac{d^{n+m}}{d\mu^{n+m}}\left(\mu^2-1\right)^n$$

The Legendre functions satisfy recurrence relationships

$$\mu\ P_n^m = \in_n^m P_{n-1}^m + \in_{n+1}^m P_{n+1}^m$$

Where $\in_n^m = \sqrt{\frac{n^2 - m^2}{4n^2 - 1}}$

The derivatives with respect to μ are determined using

$$\left(1-\mu^2\right)\frac{dP_n^m}{d\mu} = -n\mu P_n^m + (2n+1)\epsilon_n^m P_{n-1}^m$$

or

$$\left(1-\mu^2\right)\frac{dP_n^m}{d\mu} = (n+1)\epsilon_n^m P_{n-1}^m - n\epsilon_{n+1}^m P_{n+1}^m$$

3.11 Spectral Transform Method

From the grid point values $A(\lambda, \mu)$, the Fourier coefficients are determined using

$$A_m(\mu) = \frac{1}{2\pi}\int_0^{2\pi} A(\lambda,\mu)e^{-im\lambda}d\lambda$$

Spectral coefficients are calculated from the Fourier coefficients as

$$A_n^m = \frac{1}{2}\int_{-1}^{+1} A_m(\mu)P_n^m(\mu)\,d\mu$$

or

$$A_n^m = \frac{1}{4\pi}\int_{-1}^{+1}\int_0^{2\pi} A(\lambda,\mu)Y_n^{m*}\,d\lambda\,d\mu$$

These coefficients verify the Parseval-Plancherel relation

$$\frac{1}{4\pi}\int_{-1}^{+1}\int_0^{2\pi}\left[A(\lambda,\mu)\right]^2 d\lambda d\mu = \sum_{n=|m|}^{\infty}\sum_{m=-\infty}^{+\infty}\left|A_n^m\right|^2$$

The above equation indicates that global quadratic quantities such as kinetic energy can be calculated in spectral space directly.

The first coefficient of the expression A_o^o represents the mean value of the field A (λ,μ), since $Y_o^o = 1$

$$A_o^o = \frac{1}{4\pi} \int_{-1}^{+1} \int_{0}^{2\pi} [A(\lambda,\mu)] \, d\lambda \, d\mu$$

To calculate grid point values from spectral coefficients, the following two relations are used:

$$A_m(\mu) = \sum_{n=|m|}^{\infty} A_n^m P_n^m(\mu)$$

and

$$A(\lambda,\mu) = \sum_{m=-\infty}^{\infty} A_m(\mu) e^{im\lambda}$$

where $A_m(\mu)$ are the Fourier coefficients.
Transformation to and fro between Fourier coefficients and grid point values are done using Fast Fourier Transform (FFT) method (Cooley and Tuckey algorithm 1965).

For computing the spectral coefficients using Fourier coefficients, integration in y- direction (sin ϕ) is carried out using Gaussian quadrature.

$$A_n^m = \sum_{k=1}^{K} \omega(\mu_k) A_m(\mu_k) P_n^m(\mu_k)$$

where μ_k are the K roots of the Legendre Polynomial of degree K. Gaussian weights $\omega(\mu_k)$ are given by

$$\omega(\mu_k) = \frac{2(1-\mu_k^2)}{\left| K P_{K-1}^0(\mu_k) \right|^2}$$

$$\sum_{k=1}^{k} \omega_k = 2$$

μ_k are also known as Gaussian latitudes; ω_k is the weight of Gaussian latitude K; and K is the maximum number of the Gaussian latitudes.

3.12 Collocation Grid

A collocation grid is generally used for nonlinear calculations which cannot be carried out directly in the spectral coefficients. At each time step, East-West transforms are made using FFT, and Gaussian quadrature is used in the N-S directions and these Gaussian latitudes are not equidistant. At higher latitudes, lesser points are taken on a latitude circle in the Fourier mode. The grid is said to be Gaussian quadratic when the number of latitudinal circles is large enough (for a given truncation) so that Gaussian quadrature is exact for the product of two Legendre Polynomials (in the truncation).

3.13 Spectral Truncation

Since we cannot have infinite expansion, only a finite set of harmonics are used. The expressions for different types of truncation are given below and are shown in figure 3.14.

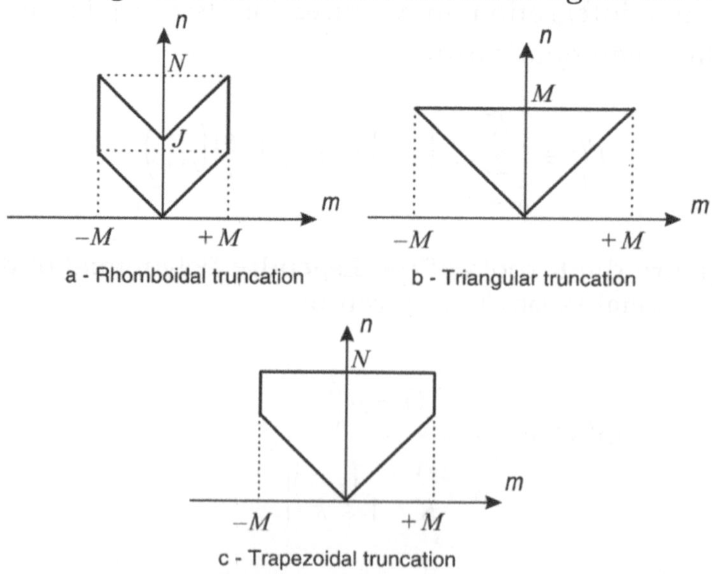

Figure 3.14 Three types of truncation

$A(\lambda,\mu)=\sum_{m=-M}^{+M}\sum_{n=|m|}^{+M}A_n^m Y_n^m(\lambda,\mu)$ represents triangular truncation and $A(\lambda,\mu)=\sum_{m=-M}^{+M}\sum_{n=|m|}^{J+M}A_n^m Y_n^m(\lambda,\mu)$ represents Rhomboidal truncation

Solution of equation

For a variable A, meridional derivative (w.r.t ϕ)

$$\left(\cos\phi\frac{\partial A}{\partial\phi}\right)_n^m=-(n-1)\in_n^m A_{n-1}^m+(n+2)\in_{n+1}^m A_{n+1}^m$$

where $\in_0^0=0$ and $\in_n^m=\sqrt{\dfrac{n^2-m^2}{4n^2-1}}$

This equation comes from the recurrence relations

$$\mu Y_n^m=\in_n^m Y_{n-1}^m+\in_{n+1}^m Y_{n+1}^m$$

and $$\left(1-\mu^2\right)\frac{\partial Y_n^m}{\partial\mu}=(n+1)\in_n^m Y_{n-1}^m-n\in_{n+1}^m Y_{n+1}^m$$

Differentiation latitudinally is done by integration by parts using zero boundary conditions at poles $\left(A^m(\pm 1)\right)=0$

$$\left[\frac{1}{a}\frac{\partial A}{\partial\mu}\right]_n^m=\int_{-1}^{+1}\frac{1}{a}\frac{\partial A^m}{\partial\mu}P_n^m(\mu)d\mu$$

$$=-\int_{-1}^{+1}\frac{1}{a}A^m\frac{d}{d\mu}P_n^m d\mu$$

$$\left[\frac{1}{a}\frac{dA}{d\mu}\right]_n^m = \sum_{j=1}^{J} A^m\left(\mu_j\right)\frac{H_n^m\left(\mu_j\right)}{a\left(1-n^2\right)}$$

where $H_n^m(\mu) = \left(1-\mu^2\right)\frac{d}{d\mu}P_{\max}$

The zonal derivative is obtained directly as $\left(\frac{\partial A}{\partial \lambda}\right)_n^m = imA_n^m$

Some useful spectral relationships for wind

$$D_n^m = \nabla^2 \chi = -\frac{n(n+1)}{a^2}\chi_n^m \quad \text{.......... Divergence}$$

$$\zeta_n^m = \nabla^2 \psi = -\frac{n(n+1)}{a^2}\psi_n^m \quad \text{.......... Vorticity}$$

$$ua\cos\phi = \frac{\partial \chi}{\partial \lambda} - \cos\phi\frac{\partial \psi}{\partial \phi}$$

$$(ua\cos\phi)_n^m = im\chi_n^m + (n-1)\in_n^m \psi_{n-1}^m - (n+2)\in_{n+1}^m \psi_{n+1}^m$$

$$va\cos\phi = \frac{\partial \psi}{\partial \lambda} + \cos\phi\frac{\partial \chi}{\partial \phi}$$

$$[va\cos\phi]_n^m = m\psi_n^m - (n-1)\in_n^m \chi_{n-1}^m + (n+2)\in_{n+1}^m \chi_{n+1}^m$$

Relationship between ζ, u and v

$$\rho = \frac{1}{a\cos\phi}\left[\frac{\partial v}{\partial \lambda} - \frac{\partial}{\partial \phi}(u\cos\phi)\right]$$

$$\rho = \frac{1}{a^2\cos^2\phi}\left[\frac{\partial(v\cos\phi)}{\partial \lambda} - \cos\phi\frac{\partial}{\partial \phi}(ua\cos\phi)\right]$$

Calculation of nonlinear terms

The nonlinear term is the product of two independent variables. To calculate the product of spectral coefficients in spectral space,

Using Triangular truncation

$$A_n^m = \frac{1}{4\pi}\int_{-1}^{+1}\int_{0}^{2\pi} A\, Y_n^{m^*}\, d\lambda\, d\mu$$

$$= \sum_{n_1 m_1}\sum_{n_2 m_2} B_{n_1}^{m_1} C_{n_2}^{m_2} \frac{1}{4\pi}\int_{-1}^{+1}\int_{0}^{2\pi} Y_{n_1}^{m_1} Y_{n_2}^{m_2} Y_n^{m^*}\, d\lambda d\mu$$

If B and C are defined with truncation 2N (T_{2N}), the product A(λ,μ) is defined with truncation 2N (T_{2N})
The spectral coefficients of A are expressed as a weighted sum of the product of coefficients B and C, the weights are constructed by the integrals, called interaction coefficients. This method is very demanding on computational time and space, as the calculation of the interaction coefficients exponentially increases with the increase of degrees of freedom for B and C.

Spectral transform method, proposed by Orszag (1970) and Eliassen et al. (1970) is far simpler and more efficient.

The following are the steps to be executed

1. The spectral coefficients of the two variables, B (λ, μ) and C (λ, μ) [B_n^m and C_n^m] of the product are first transformed on to spatial grid G individually

   G= [λ_j, j=1,2,3,....................j], [μ_k, k=1,2,3,.....................K] to yield B (λ, μ) and C (λ, μ).

2. The values of B (λ, μ) and C (λ, μ) are multiplied at the grid points,

   A (λ, μ) = B (λ, μ) * C (λ, μ)

3. As next step, the grid point values of A (λ, μ) are transformed to the spectral space as the spectral coefficients , A_n^m using

$$A_n^m = \frac{1}{2}\int_{-1}^{+1}\left[\frac{1}{2\pi}\int_0^{2\pi} A(\lambda,\mu)e^{-im\lambda}\,d\lambda\right]P_n^m(\mu)\,d\mu$$

4. The step 3 above is carried out as two steps again to compute the two integrals, first for the calculation of Fourier coefficients and then the spectral coefficients.

5. The Fourier coefficients (for each of the latitudinal circle) are calculated from the grid point values using

$$A_m(\mu_k) = \frac{1}{2\pi}\int_0^{2\pi} A(\lambda,\mu)\,e^{-im\lambda}\,d\lambda$$

This integral is calculated using trapezoidal quadrature as

$$A_m(\mu_k) = \frac{1}{L}\sum_{j=1}^{L} A(\lambda,\mu)\,e^{-im\lambda}$$

This finite sum on J regularly spaced points yields an exact result of the integral if its integrated sum is a sum of sine functions of a degree less than J-1. In the case of product A=B * C, the integrand is written as a sum of sine functions of a degree less than or equal to 2M+M=3M. So an exact evaluation of integral is achieved by choosing J >= 3M+1.

6. Then the second integral is calculated

$A_n^m = \frac{1}{2}\int_{-1}^{+1} A_m(\mu)P_n^m(\mu)\,d\mu$ using Gaussian numerical quadrature.

$$A_n^m = \frac{1}{2}\sum_{k=1}^{k} W_k A_m(\mu_k) P_n^m(\mu_k)$$

7. The transformation at step 6, is performed for different values of μ_k (sin θ_k), k=1,2,..........K, corresponding to different (Gaussian) latitudes using weights

   $$W_k = \frac{2\left(1-\mu^2\right)}{\left[k\, P_{k-1}^0\left(\mu_k\right)\right]^2}$$

   This weighted sum on K points yields an exact result when the integral is a polynomial of degree less than or equal to 2K-1.

In the case of the product A=B*C, exact integral is obtained by choosing K >= 3M+1

In the case of triangular truncation

J >= 3M+1
K >= 3M+1/2 (since 2K-1 >= 3M)

To sum up the procedure,

1. Use FFT (Fast Fourier Transform) to transform A (λ, μ) values, at each of the latitudes, to Fourier coefficient A^m (μ) through trapezoidal quadrature.

2. Use Legendre transform to transform A^m (μ) to spectral coefficients A_n^m using Gaussian quadrature.

3. Use inverse Legendre transform by summation over the total node number 'n' (i.e.) the transformation of spectral coefficients A_n^m to the Fourier coefficients A^m (μ) on Gaussian latitudes

   $$A_m(\mu) = \sum_{n=|m|}^{M} A_n^m P_n^m(\mu)$$

4. Use inverse FFT , by summation over zonal wave number 'm', to transform the Fourier coefficients to grid point values A (λ, μ)

$$A(\lambda,\mu)=\sum_{m=-M}^{M} A_m(\mu)e^{im\lambda}$$

Some important observations

1. Spectral method has decisive advantage over the finite difference method. In the finite difference method, the truncation errors are larger for higher wave numbers (smaller scales). Spectral method provides higher accuracy due to series expansion for all the wave numbers in the truncation.

2. Since the grid distance becomes smaller as we move from equator to pole, the contributing terms from highest wave numbers becomes extremely small, and so considering less number of points at higher latitudes saves computational time.

3. Choice of truncation

4. Computation of quadratic terms without aliasing.

Part D. Vertical Coordinate Systems

A proper delineation of the vertical structure of the atmosphere by choosing appropriate vertical coordinate is necessary as of the horizontal coordinate. All models use discretized vertical structure unlike horizontal which may be represented through grid point or spectral methods. The forecasts, thus produced represent layer-averaged values at the different levels. Herein, few types of vertical coordinate systems are presented.

Most hydrostatic models use relatively straightforward configurations for their vertical coordinates. However, no one vertical coordinate system is ideal; each has its own strengths and limitations.

3.14 Different Vertical Coordinate Systems

3.14.1 Pressure Coordinate

In meteorology, the hydrostatic equation shows the relationship between height and pressure in the vertical column indicating the possibility of using pressure as vertical coordinate. A distinct advantage of the isobaric coordinate system is that the horizontal pressure gradient force is measured by the gradient of geopotential at constant pressure, and so density does not explicitly appear in the calculation of pressure gradient force. Another advantage is that the equation of continuity is linear. However, the disadvantages are that the lowest model pressure surface does not coincide with the earth surface making it necessary to make adjustments to the computation of vertical velocity in sloping terrain. The assumption of constant pressure at the lower boundary also leads to errors in the estimation of near surface meteorological variables.

3.14.2 Sigma Coordinate

The sigma coordinate is defined by $\sigma = p/ps$, where p is the pressure on a forecast level within the model and ps is the pressure at the earth's surface (not mean sea level pressure). The lowest coordinate surface follows the actual terrain. The terrain slopes in sigma models are always smoothed to some extent, and the other sigma surfaces gradually change from being nearly parallel to the terrain at the bottom of the model ($\sigma = 1$) to being nearly horizontal to the constant pressure surface at the top of the model ($\sigma = 0$).

Advantages of this system are : (i) terrain following coordinates allows good representation of atmospheric fields such as wind and temperature where terrain variations are smooth and wide, (ii) better resolution near the ground all over the domain and (iii) coordinate surfaces do not intersect with the earth surface unlike other coordinates such as pressure or isentropic coordinates.

The limitations are : (i) Computation of pressure gradient force involves use of lapse rates, which will induce errors in the regions of steep terrain. This may lead to large errors in the derivation of wind field. (ii) Smoothening of the steep terrain will misrepresent the local terrain that leads to errors in the derivation of pressure, temperature and moisture at the surface. (iii) Due to smoothening of terrain near the coastal regions, land points may be forcibly moved beyond the coastline. (iii) weather systems in the lee of mountain regions may be distorted (such as excessive leeside cyclogenesis due to enhanced vortex-tube stretching).

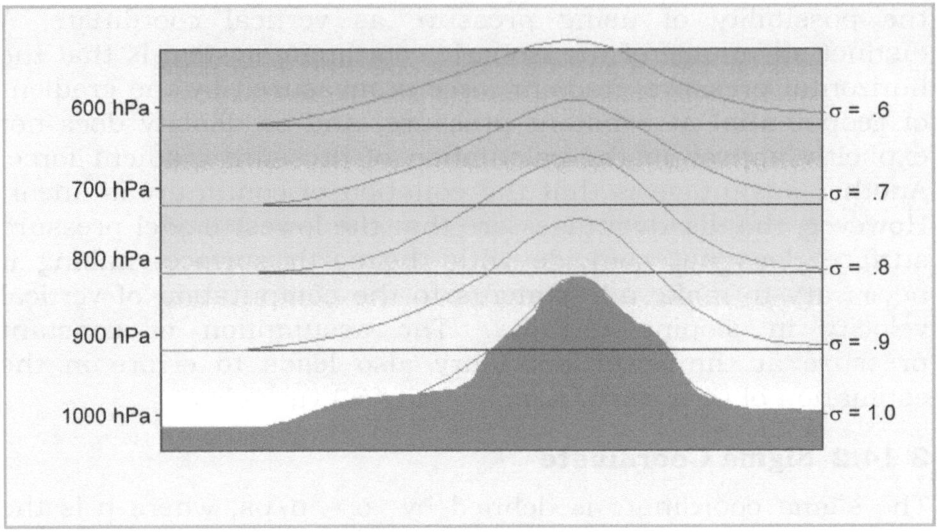

Figure 3.15 Representation of Sigma vertical coordinate surfaces (from COMET)

3.14.3 Eta Coordinate

The Eta coordinate is another form of the sigma coordinate, in which mean sea level pressure is used in the place of surface pressure as the bottom reference level. The use of the Eta rather than sigma coordinate improves the calculation of winds and vertical motion in areas of steep terrain. As such, Eta is defined as

$$\eta = [p_r(z_s) - p_t] / p_r(z = 0) - p_t]$$

where z is geometric height, p_t is the pressure at the top of the model, p_r (z = 0) is the mean sea level pressure (1013 hPa), and p_r (z_s) is the standard atmospheric pressure at the model terrain level z_s.

Eta is labelled from the top of the model domain to mean sea level, as 0 at the top and 1 at the mean sea level. Unlike sigma coordinates where all the grid points lie above the earth's surface, in eta system some grid points may be located below the surface elevation in areas where the elevation is significantly above sea level.

The bottom layer of the model within each grid box is represented as a flat "step," rather than as a sloping sigma over steep terrain. For this reason, the Eta coordinate is sometimes referred to as the step-mountain coordinate. This configuration eliminates most of the errors in the calculation of pressure gradient force term but may lead to large differences in elevation from one grid point to its neighbouring point. Therefore, Eta coordinate models can produce strong vertical motions in areas of steep terrain to have more accurate representation over the mountains. Even with the step-like Eta vertical coordinate, model terrain can still be much coarser than real terrain, but the topographic gradients will be less smooth than in sigma coordinates. Although this can be a source of error in areas of small-scale terrain features, representation of the average elevation for the entire grid box area makes the forecast to be better representative of the average conditions of the grid box area.

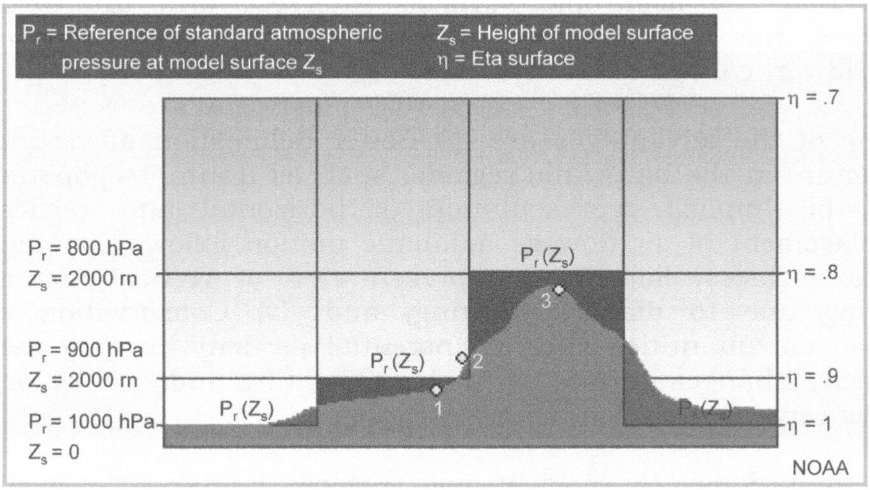

Figure 3.16 Eta vertical coordinate system (from COMET)

The Eta coordinates have some significant advantages, because of the near horizontal coordinate surface and with better representation of topography. Some of the advantages are: (i) Minimal errors in the calculation of pressure gradient force term due to reduction in vertical interpolations as in sigma coordinates which would improve the prediction of wind, temperature and moisture in the regions of sloping terrain. (ii) Better prediction of low level convergence in the areas of mountains resulting in better prediction of precipitation in the mountainous regions. (iii) Better prediction of temperature gradients in systems such as lee cyclogenesis, cold fronts etc.

The Eta coordinate system has some disadvantages also, which are: (i) Inadequacies in the representation of the vertical structure in boundary layer over steep mountainous terrain, (ii) Inaccurate description of gradually sloping terrain over large distances, (iii) Difficulties in the prediction of downslope winds, (iv) Inadequate representation of valleys covering a few of the grid boxes, and (v) generation of spurious waves at the grid box edges, which may have considerable impact if the horizontal resolution is finer than 10 km and the vertical resolution is coarse.

3.14.4 Isentropic (theta) Vertical Coordinate

Potential temperature is taken as the vertical coordinate, considering the free atmosphere to be mostly isentropic. Since non-adiabatic processes predominate in the boundary layer and also that the isentropic surfaces intersect with the earth surface, this system is not directly used but considered in hybrid vertical coordinate systems.

Some of the advantages are: (i) Better delineation of vertical structure in the baroclinic regions, such as fronts, tropopause etc., (ii) Implicit representation of horizontal and vertical displacement of air flow as adiabatic motion follows constant theta surfaces, (iii) Better representation of vertical motion arising due to diabatic heating, and (iv) Conservation of dynamical quantities such as potential vorticity, as potential vorticity changes caused by diabatic heating and cooling are better represented.

The limitations are: (i) Isentropic surfaces intersect the earth surface, and for this reason isentropic coordinates are used in combination with sigma coordinates near the ground (as in hybrid coordinates), (ii) Isentropic coordinates in the boundary layer may show superadiabatic layers which constrain prediction of some weather systems, and (iii) Vertical resolution will be coarse in adiabatic layers leading to improper representation of vertical mixing in such adiabatic layers.

3.14.5 Hybrid Vertical Coordinate

Since there are advantages and limitations with each of the coordinate systems described above, hybrid coordinate systems such as sigma-theta and sigma-pressure coordinates have been developed and experimented.

3.14.5a. Hybrid isentropic-sigma coordinate: This system has a combination of sigma layers at the bottom slowly transiting to isentropic layers above. This retains the advantages of terrain-following sigma coordinates near the surface and the increased vertical resolution in important baroclinic regions at higher levels. It is important that the boundary sigma layer is deep enough (at least 200 hPa) to resolve the diurnal variations of the boundary layer processes and to avoid formation of superadiabatic layers.

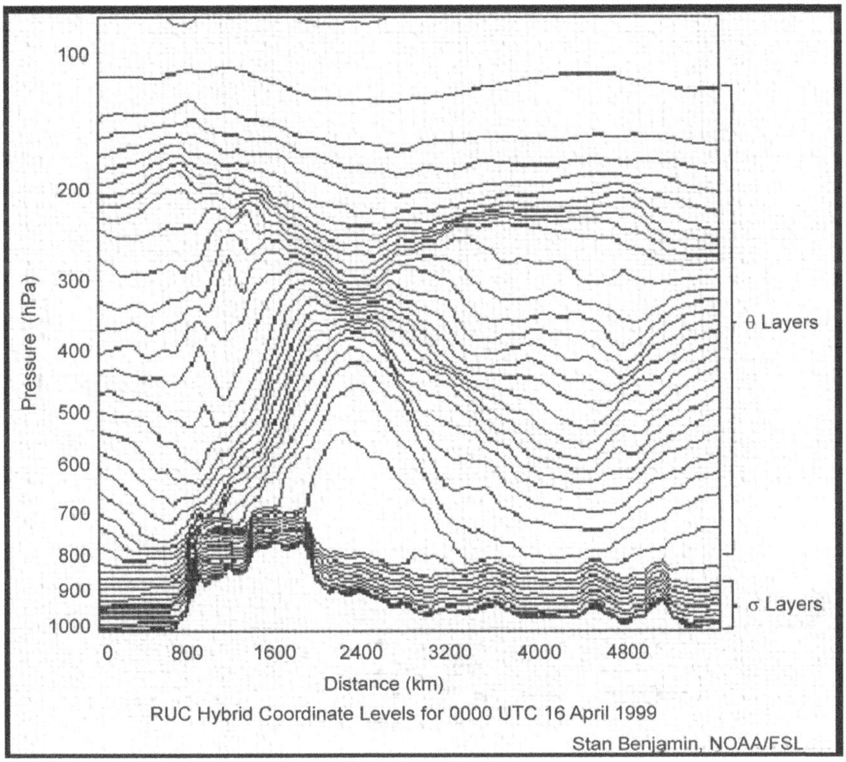

Figure 3.17 Hybrid sigma-theta vertical coordinate system

The advantages are: (i) Isentropic system at higher levels resolves warm advection better than sigma coordinates leading to better prediction of precipitation, (ii) no problem of isentropic surfaces intersecting the surface, (iii) good representation of surface heating and mixing in the boundary layer, and (iv) better representation of surface physics interactions such as surface evaporation.

The limitations are: (i) No possibility of adiabatic flow representation at lower levels, (ii) Exact depiction of processes near the boundary layer-free atmosphere interface due to non-concurrence of sigma layers with the depth of the boundary layer, and (iii) difficulties to blend the two coordinate types at their interfaces.

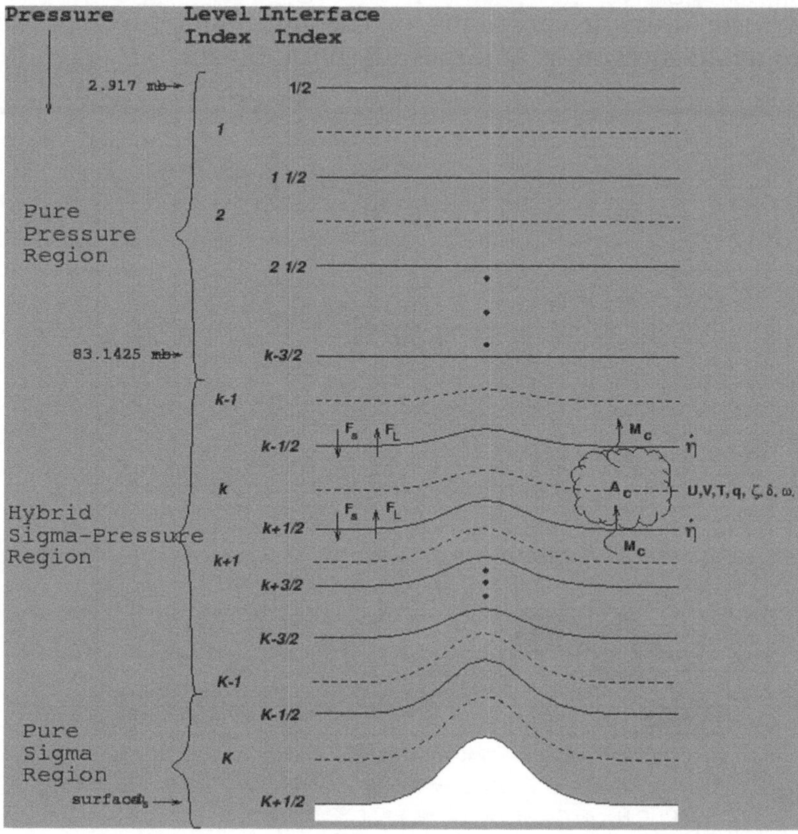

Figure 3.18 Hybrid sigma-pressure vertical coordinate system

3.14.6 Hybrid Sigma-pressure Coordinate System

This coordinate system has sigma layers at the bottom with transition to isobaric surfaces at higher levels. This system keeps the advantages of terrain-following sigma in the boundary layer and near surface and smoother and planar surfaces at higher levels which facilitate the assimilation of radiances and calculations of radiative transfer. This system may be having terrain following surfaces to higher levels in the mountainous regions and planar horizontal surfaces in the remaining areas. This also has the advantage of reducing the computational error of the pressure gradient force on sloping surfaces. However, difficulties arise at the interface between the two surfaces. The computations in the hybrid coordinates take longer time than the single vertical coordinate systems.

Part E. Boundary Conditions

In NWP, use of limited area models is important as short-range weather predictions are desired at as high horizontal resolution as possible for variety of applications. In recent times, apart from mesoscale weather predictions, weather information at very high resolution is required for application in agriculture, wind energy, air quality assessment, urban hydrology etc. Since global models cannot be integrated at the desired high resolutions and atmospheric changes over short time scales are controlled by regional synoptic regimes, it may suffice to use limited area models for certain of the identified applications. However, limited area models need proper specification of the lateral boundary conditions through the time integration period. In this section, methods for the determination of the boundary conditions and the current strategies are described.

Limited area models (LAM) are first defined by their horizontal coverage of area, referred to as domain. Global models, covering the entire horizontal extent of the globe need specification of information at vertical boundaries (i.e.) at the surface and top of the (model) atmosphere whereas LAMs additionally require lateral boundary conditions.

The model domain could be envisaged as a 3-dimensional array of cubes, with each of the cube covering a fixed volume of the atmosphere. Each side of the cube receives information from neighbouring cubes except those situated at the boundaries. The sides which are exposed to lateral sides need specification of information from external sources.

3.14.7 Lateral Boundary Conditions

One way is to derive the lateral boundary conditions for the LAMs from the output of a coarser model of larger domain or mathematically impose the conditions at the lateral boundaries. It is important to note that the values of all the forecast variables at the boundaries should be as accurate as possible. The lateral boundary conditions may be determined following one of the procedures: (i) choosing climatological fixed values for certain variables such as soil moisture, sea surface temperature, surface characteristics (type such as vegetation); (ii) use of forecast values from a coarser model (needs interpolation to model domain); and (iii) data assimilation system.

While determining the values at the boundaries, some of the concerns are: (i) the model area nearer to the boundaries will have higher impact of the boundary information taken from another source as errors of the model prediction would propagate inward, (ii) the deficiencies of the coarser model forecast will impact as amplification of errors in the model forecast, and (iii) the lower boundary conditions will broadly impact the physical parameterizations that would reflect in the temperature, wind and precipitation fields.

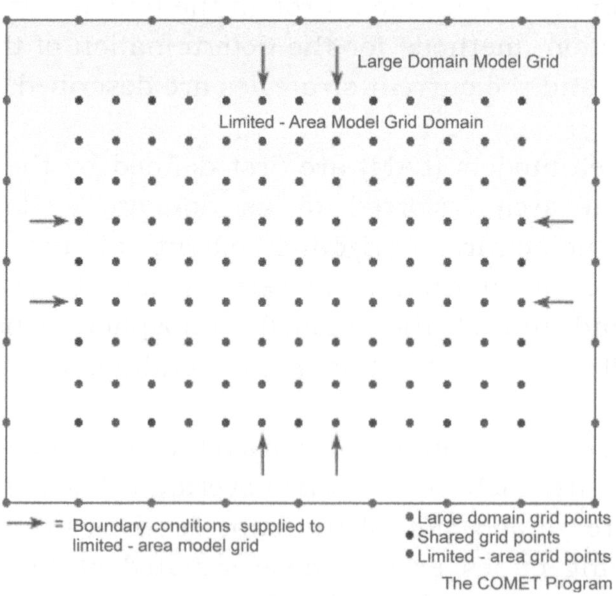

Figure 3.19 Limited area model domain (from COMET)

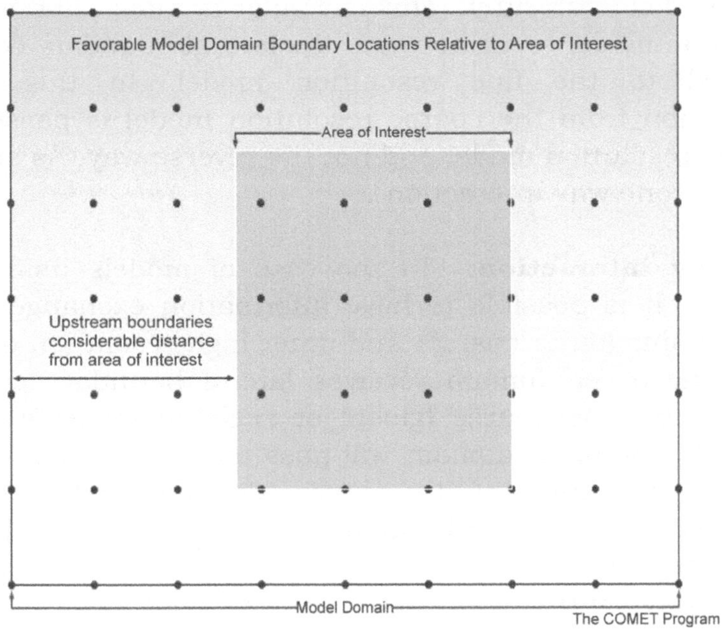

Figure 3.20 Model domain and boundaries (from COMET)

The lateral boundary conditions will influence the development of the weather patterns over the forecast domain, smaller the domain size faster will be the impact. For e.g., the forecast of a synoptic system in a high resolution model would be controlled by the features in the coarser model due to impact of boundary conditions, due to the spread of boundary values downstream. It is suggested that proper assessment is to be made about the air movement from the boundaries to the forecast domain of interest and grid domain considered accordingly.

At the present time, limited area models are designed to have nesting, which means that the model can have different domains with different resolutions embedded together. The outer domain will have coarser resolution covering a larger area and the inner domain will have higher resolution of smaller area placed within the outer domain. In this case there are two ways in which the domains interact with each other.

One way interaction: The inner domain would be receiving boundary conditions from the outer (mother) domain through the model integration time, as both domains would be integrated concurrently. This is similar to integration of coarser resolution model separately and the lateral boundary conditions provided to the fine resolution model. In this method, information from the coarse resolution model is passed on to the fine resolution model and not the reverse way. As such, this is called "one way interaction".

Two way interaction: In the case of models using nested domains, it is possible to have information exchange between two domains (outer coarser and inner high resolution domains). While the outer domain receives lateral boundary conditions from an outside source (global or regional model with wider domain), the outer domain will pass on the information at the lateral boundaries of the inner domain and information generated at the lateral boundaries of the inner domain will be passed on to the outer domain. Since information is exchanged in both ways, it is called "two way interaction".

Due to the use of lateral boundary conditions taken from a coarser model, errors may result due to the following: (i) accuracy of the model forecast providing the boundary conditions, as the boundary information may not represent gradients suitable for the finer grid and inability of the coarser model to include the small resolvable scale features of the fine resolution model, (ii) differences in the parameterizations of the physical processes between the coarser and finer resolution models cause forecast errors, and (iii) differences in vertical resolution, topography at the boundaries and also the dynamical inconsistencies arising due to different vertical coordinates. In addition, the propagation of waves in the coarser model at the boundary interface with the fine resolution domain may be transformed in the characteristics and sometimes may be reflected at the boundary or refracted through its passage within the inner domain. All these errors will affect the prediction of vertical motion, temperature and precipitation fields.

Upper boundary conditions: All types of models (global, regional and LAM) need specification of values at the top and bottom boundaries of the model domain. Generally, the top of the model atmosphere is located near or above the tropopause and clear assumptions are to be made about the change of the forecast variables through model integration time. It is simpler to use "rigid" upper boundary condition, meaning that no vertical motion is allowed to pass through the top. This assumption is reasonable most of the times, except when gravity waves generated by convection or orography are present near the top levels which will be subject to reflection due to rigid boundary assumption. If proper care is not taken, these reflected gravity waves will impact the vertical motion fields at lower levels and consequentially the precipitation forecast. To alleviate this, an "absorbing" or "damping" layer is to be created at the highest model layers and the model top to be taken much higher than the vertical scale of the weather systems to be predicted.

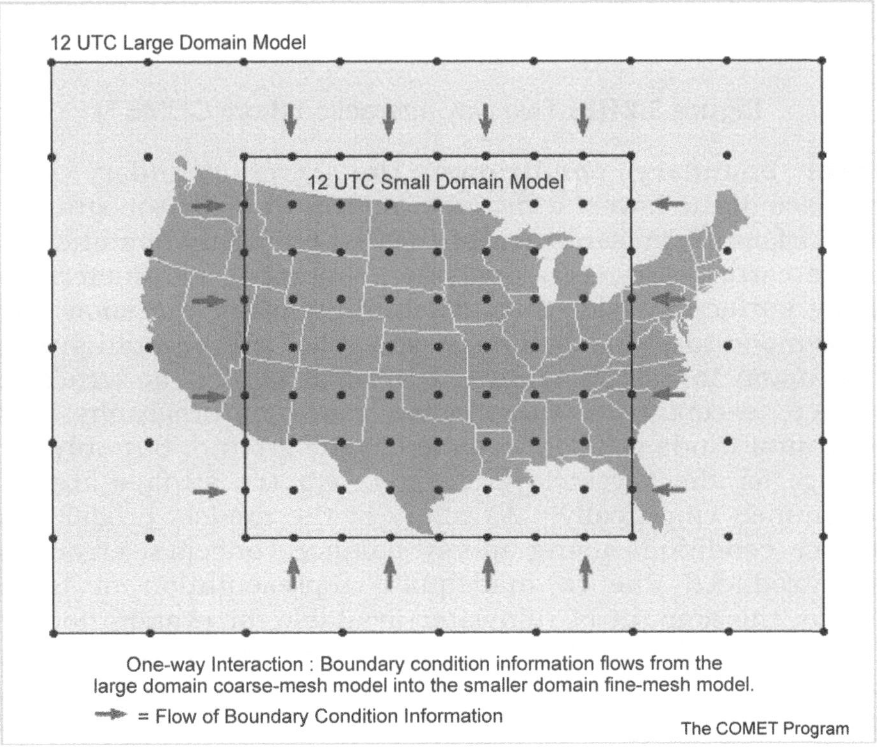

Figure 3.21(a) One way interaction (from COMET)

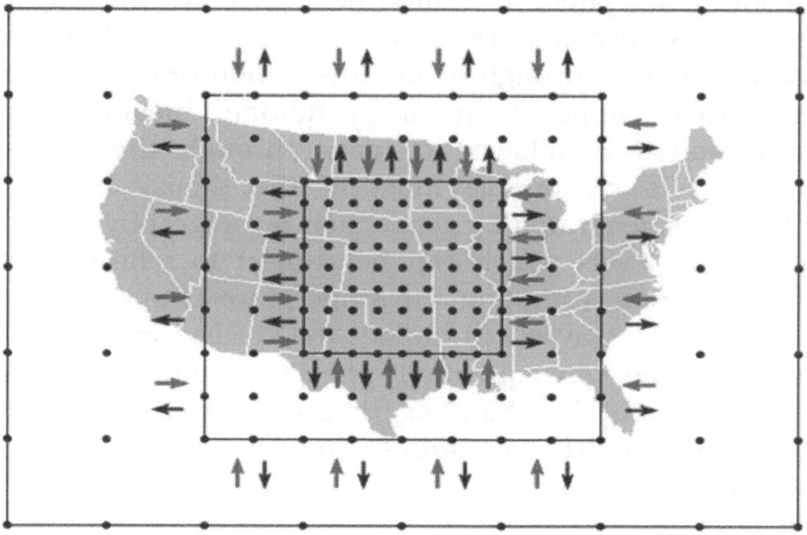

Figure 3.21(b) Two way interaction (from COMET)

Lower boundary conditions: The lower boundary is the interface of the lowest atmospheric level with the topography or sea surface. Representation of the low boundary condition (i.e.) at the earth's surface is dependent upon the parameterization of the surface physics and the information of the snow cover, soil temperature and moisture, soil type, and vegetation cover. It is usual to assume that the vertical motion is zero at the surface, except over the sloping parts of orography. Since horizontal winds are not predicted at the ground, but only as an average for the lowest layer, winds near the surface are to be determined empirically. As many of the models predict near-surface conditions using energy balance concepts, errors may be introduced due to inadequate representation of terrain, albedo, the amount of rainwater available for evaporation from the surface, lake and sea temperature, vegetation cover, the method of simulating soil-vegetation-atmosphere interaction, and many other details related to representation of physical processes.

Since models use defined surface characteristics as part of their bottom boundary conditions, errors can be introduced because of: (i) the surface not being represented with enough resolution to generate a good forecast, (ii) use of model data (not observations) to define surface conditions that lead to biases and errors in surface conditions, (iii) model specifications may not represent the actual surface conditions and so will affect the local forecast, and (iv) inadequate representation of the processes that influence the model forecast.

It is important to have prior knowledge of model deficiencies, as surface processes have a profound impact on sensible weather at the ground and on local weather forecasts and inadequate representation of these processes contributes to forecast error.

4. Objective Analysis

For the purpose of numerical weather prediction, atmospheric models are used. As described in the previous chapters, the models consist of mathematical equations governing the atmospheric motion, which are solved using numerical methods. Two numerical methods, one is based on finite difference approximations and the other on the spectral method, are used for solution. These two methods are described separately in chapter 3 of this book. In the finite difference methods, all the model variables are to be provided at the model grid points in the 3-dimensional space. Atmospheric data are conventionally collected all over the globe at the specified observation sites and at prefixed synoptic times. Since these observation stations are positioned at selected but irregularly spaced sites, the data values are to be redefined on the model grid.

In 1940s and 1950s, the initial values on the model grids were produced by subjectively plotting the data on charts, performing analysis of drawing isolines and then pick up the value at the grid point through subjective interpolation between isolines. This was a slow and laborious process. Later objective analysis (OA) methods were developed and applied to determine the required model grid information. The OA method was found to be much faster and accurate than the subjective method.
The objective analysis methods described here are confined to spatial interpolation.

4.1 Function Fitting Method

In this method, the observations in an area are fit to mathematical equation, such as a polynomial equation (Panofsky, 1949). A third degree polynomial can be written as

$$p\,(x, y) = \Sigma_{i,j}\, a_{ij}\, x^i\, y^j \qquad (i + j \leq 3)$$

where x and y are Cartesian coordinates and a_{ij} are the ten coefficients. The size of the area to which the polynomial is to be fitted will depend on how many of the observations are to be smoothed. A third degree polynomial (shown above) will exactly fit 10 observations without any smoothing. The coefficients are obtained by using the method of least squares where the number of observations is more than the coefficients. The solution of a cubic polynomial yields to 10 normal equations, which are to be solved as simultaneous equations. The procedure generally takes considerable computing time.

4.2 Empirical Linear Interpolation

In this method, model grid point values are derived using the values of the variable from nearby locations.

The basic form of the linear equation is:

$$X_{grid} = \sum_{1,m} h_j X_{oj}$$

where X_{grid} is the interpolated grid point value; h_j is the scalar weight at point j; X_{oj} is the observation value at point j; and m is the number of observations.

The scalar weight h_j is of the form:

$$h_j = W_j \Big/ \sum_{j=1,\, m} W_j$$

where w_j is the weighting function applied to point j, and that the sum of all h_j equals one $[\sum(h_j)=1]$.

The weight function (w_j) can be determined in a number of ways, of which one method is to use empirically derived functions. The empirically derived weight functions use distance proportional weighting in which observations that are closer to the grid point are given more weightage than that are farther away.

In principle, all available observations can be used to determine the value at a grid point. However, since the weighting function decreases with distance (i.e.) observations farther away may have little influence and so to save computational time, a threshold distance called the "radius of influence" is adopted in the calculation. This radius of influence indicates the degree of smoothing, smaller radius meaning lesser smoothing and larger radius meaning higher smoothing. Generally the properties of the weighing functions are such that closer observations will have higher influence, unit weight for a coincident observation, and resulting field is differentiable. Since the weighting function is generally determined by trial and error, the method is empirical.

4.3 Successive Correction Method

In this method, a first guess field (assigned values at all grid points) is considered, preferably an earlier short range weather forecast from NWP. This is an iterative process, in which the first guess field is modified using a linear weighting scheme to fit the grid point values to the observations repeatedly till the errors are within a specified tolerance limit.

In the empirical linear interpolation method, actual observations are used to calculate grid point values. The successive corrections approach begins with a first guess or background field and then the differences between the background values and the observations are used to obtain the final grid point values through an iterative process. This method is useful for grid areas where observations are sparse. The following step-wise procedure is followed in this method.

Estimate the background value at each observation point [X_{bj}] using the first guess values.

Calculate the difference between the background value and the corresponding observation value [$X_{oj} - X_{bj}$] at each observation location. This is called prediction error.

Calculate the prediction errors at each of the grid points using linear interpolation.

Correct the grid point values based on the distributed errors. [$X_{grid(k+1)} = X_{grid(k)} + \sum_{j=1 \text{ to } m} (h_j * \{X_{oj} - X_{bj}\})$], where k is the iteration step index.

Repeat the above steps until the value of all predicted errors is below a pre-defined value.

This iterative process will forge the grid point values to converge to the observation values. At each iteration step, the weights or the radius of influence can be modified.

4.4 Cressman Scheme

The Cressman (1959) scheme, described here, is the first operational successive corrections OA scheme used at the National Weather Service, USA during the early period of numerical weather prediction.

It is similar to the successive corrections scheme. It uses a non-zero weighting function within a prescribed radius of influence. This radius of influence decreases with each successive scan. After a large number of scans, the grid values converge to the observations. In areas of high observations density, the grid point values reflect the observations, while in low-density areas the grid values remain closer to the first guess or background field.

The Cressman weighting function takes the form:

$$W_j = \frac{(d^2 - r^2)}{(d^2 + r^2)} \quad \text{for } r < d$$

$$w_j = 0 \quad \text{for } r \geq d$$

where r is the distance from the grid point to the observation, d is the radius of influence.

This original weighting function was modified slightly as

$$w_j = [(d^2 - r^2) / (d^2 + r^2)]^{\alpha} \qquad r < d$$

$$w_j = 0 \qquad r \geq d$$

where $\alpha > 1$.

Assignment of values for d and α is empirical and depends upon the data spacing and the desired level of smoothing. In general, a smaller radius of influence has less smoothing; a larger radius of influence produces a higher degree of smoothing. For the original Cressman scheme, a value for d as twice the average spacing of the observations was adopted to keep a balance between under-smoothing and over-smoothing.

4.5 Barnes Scheme

Barnes (1964) proposed a scheme to interpolate randomly spaced data onto a grid with a desired level of accuracy. The Barnes weighting function is given by:

$$w_j = \exp\left[-r^2 / 4k \right]$$

where 'r' is the distance from the grid point to the observation, and 'k' is the parameter to define the response of the weighting function. The value of k is chosen so that when r = d (d is the radius of influence), the weighting function is e^{-4} times maximum value of data at r = 0. The radius of influence is chosen so that it is greater than the average spacing of the observations.

The iterative process for obtaining the grid point values is as follows:

On the first scan, estimate the value of the variable at each grid point using the following equation

$$x_{i,j} = \left[\left(\sum_{n=1 \text{ to } N} w(r,d) * X_{on}\right) / \sum w(r,d) \right]$$

where $x_{i,j}$ is grid point value at location i,j; X_{on} is the observed value; N is the number of observations; and w(r,d) is the Barnes weight function.

Next, estimate the variable value for each observation location by averaging the four closest grid values.

Calculate the difference between the estimated value and the observation value (residual) at each of observation location.

Distribute the differences on to the grid points using the linear interpolation scheme described above with the Barnes weighting function.

Repeat the last three steps (2 to 4) until the residual is less than a pre-defined value.

4.6 Optimum Interpolation Method

In this method, the ensemble average of the squared difference between the gridded and observed data is minimized using the method of least-squares. In this approach also, a first guess (background) field, usually taken from a model forecast, is modified blending all available observations. As a first step, values of the model fields are derived at each of the observation location using an objective analysis procedure. The differences between the model based values and the actual observation values, which are called observation increments, are obtained at each observation location. The spatial distribution of the observation increments (i.e. at each of the model grid point) are to be derived. Since these observation increments are estimates of the changes in the background field, the next step is to find out the weights that are to be assigned to each of the observation location in order to estimate the observation increment value at each of the model grid point. This means that the distribution of weights would be different for each of the variable and for each of the model grid point. Then, the analysed increments are added to the background values at each model grid point to generate a new estimate of the variable.

In mathematical form,

$$F_k^a = F_k^p + \sum_{i=1}^{N} w_{ik} f_i^o \qquad .....(4.1)$$

where F represents a full-field variable and f represents an incremental variable. The superscripts a, p, and o denote the analysed, predicted (background), and observed values respectively. The subscripts i and k represent the observation and model grid point (analysis) locations respectively. The w_{ik} is the weight, related to the observed increment f_i^o for the analysis at grid point k. The novelty of this method is the derivation of weights w_{ik} through optimization.

The optimum interpolation (OI) procedure determines the weights such that the mean square error of the analysis is minimized when calculated over a statistically significant number of cases. It is important to note that limitations arise due to the assumption and errors in the determination of the statistical parameters.

Let F_k^t is the true value of a variable at a given location. Although our interest is to determine the value of F_k^t exactly, it is only possible to make an estimate considering all available information regarding the variable. The first guess value (taken from the short range forecast) F_k^p is one such estimate, as it contains past information of the variable(s) consistent with the model physics and dynamics. The observations F_i^o represent the existing atmospheric state. The objective analysis scheme, using the information from the first guess field and observations, produces another estimate F_k^a, which is another estimate of F_k^t . None of these estimates may exactly represent F_k^t, as each of them has its own sources of error. Thus, we can write the error associated with each individual value as

Analysis Error: $E^a = F_k^a - F_k^t$ .....(4.2)

Observation Error: $E^o = F_i^o - F_i^t$ .....(4.3)

Prediction Error: $E^p = F_k^p - F_k^t$ .....(4.4)

By assuming that the value (or mean) of these errors is zero considering a large ensemble of samples, their statistical error estimates can be obtained as

Analysis Error Estimate $E^a = \left\langle (E^a)^2 \right\rangle^{1/2}$ .....(4.5)

Observation Error Estimate $E^o = \left\langle (E^o)^2 \right\rangle^{1/2}$ .....(4.6)

Prediction Error Estimate $E^p = \left\langle (E^p)^2 \right\rangle^{1/2}$ .....(4.7)

where the angle brackets indicate the ensemble average. Normalising the various errors by the respective error estimates, we have

$$\varepsilon^p = E^p / \overrightarrow{E^p} \quad .....(4.8)$$

$$\varepsilon^0 = E^0 / \overrightarrow{E^0} \quad .....(4.9)$$

$$\varepsilon^a = E^a / \overrightarrow{E^a} \quad .....(4.10)$$

Similarly, the observation and analysis error estimates are normalized by the prediction error estimate giving

$$e^0 = E^0 / \overrightarrow{E^p} \quad .....(4.11)$$

$$e^a = E^a / \overrightarrow{E^p} \quad .....(4.12)$$

The observed and the analysis increments will be represented in dimensionless form as

$$f_i^0 = \frac{(F_i^0 - F_i^p)}{\overrightarrow{E^p}} \quad .....(4.13)$$

$$f_k^a = \frac{(F_k^a - F_k^p)}{\overrightarrow{E^p}} \quad .....(4.14)$$

The analysis at a specific point k, for a set of N observations, could be expressed as shown in equation (4.1). The analysed increment f_k^a at point k is given by the summation

$$f_k^a = \sum_{i=1}^{N} w_{ik} f_i^o \quad .....(4.15)$$

By substituting equations (4.13) and (4.14) into equation (4.1), squaring, and taking the ensemble average, an expression is derived for the error using the analysis at location k and considering the relationships between the normalized prediction and observation errors.

After some arithmetic manipulations, the normalized analysis error variance at location k as

$$(e_k^a)^2 = 1 - 2\sum_{i=1}^{N} w_{ik}\ h_{ik} + \sum_{i=1}^{N} \sum_{i=1}^{N} w_{ik}\ w_{jk}\ M_{ij} \quad .....(4.16)$$

where

$$h_{ik} = \left\langle \varepsilon_k^p\ \varepsilon_i^p \right\rangle - \left\langle \varepsilon_k^p\ \varepsilon_i^o \right\rangle e_i^o \quad .....(4.17)$$

$$Mik = \left\langle \varepsilon_k^p \ \varepsilon_i^p \right\rangle + e_i^o \left\langle \varepsilon_i^o \ \varepsilon_i^o \right\rangle e_j^o - e_i^o \left\langle \varepsilon_i^o \ \varepsilon_j^p \right\rangle - \left\langle \varepsilon_i^p \ \varepsilon_j^o \right\rangle e_j^o \quad .....(4.18)$$

The terms inside the angled brackets are error correlations. For example, $\left\langle \varepsilon_i^p \ \varepsilon_j^p \right\rangle$ and $\left\langle \varepsilon_i^o \ \varepsilon_j^o \right\rangle$ designate the prediction and observation error correlations respectively between the observation locations i and j. In practice, the observation and prediction errors are assumed to be uncorrelated, which means that the terms containing cross-correlations [such as $\left\langle \varepsilon_i^p \varepsilon_j^o \right\rangle$ in equations (17) and (18)] are equal to zero. With the assumption that there is no correlation between the errors of observations made at different locations, the observation error correlations [such as $\left\langle \varepsilon_i^p \ \varepsilon_j^o \right\rangle$] will be equal to either one or zero for i = j or $i \neq j$ respectively.

These assumptions reduce the expressions for h_{ik} and M_{ij} as

$$h_{ik} = \left\langle \varepsilon_k^p \ \varepsilon_i^p \right\rangle \quad .....(4.19)$$

$$M_{ik} = \left\langle \varepsilon_i^p \ \varepsilon_j^p \right\rangle + e_i^o \left\langle \varepsilon_i^p \ \varepsilon_j^o \right\rangle e_j^o \quad .....(4.20)$$

The computation of M_{ij} for all of the i, j pairs yields an N x N matrix M, in which the diagonal term on row i is equal to 1 + $(\varepsilon_i^o)^2$ and the off-diagonal terms on row i are equal to $\left\langle \varepsilon_i^p \varepsilon_j^p \right\rangle$, which are the prediction error correlations between location i and the other N - 1 observation locations. Analogously, the calculation of h_{ik} for all observations yields an N-dimensional row vector h_{ik} designating the prediction error correlation between the N observation locations i and the analysis grid point location k.

The essence of this Optimum Interpolation (OI) methodology is to determine the weights w_{ik}, which minimizes the analysis error variance in equation (4.16). As with any minimization problem, equation (4.16) is subjected to partial differentiation with respect to w_{ik} for i = 1, ...N; equate the derivatives to zero; and solve the emerging set of N linear equations for the weights.

The solution for this problem is

$$w_k = M^{-1} h_k \qquad .....(4.21)$$

where w_k represent the N-dimensional row vector of weights w_{ik} and M^{-1} is the inverse of matrix M. This means that , the weight gotten by observation i for the analysis at grid point k is a function of (i) the normalized observation error, (ii) the normalized prediction error correlations between the observation location i and the other N-1 observation locations, and (iii) the normalized prediction error correlations between the N observation locations and the grid location k. From this deduction, it can be inferred that the computed weights for the observations are dependent upon the designation of the observation error estimate $\overrightarrow{E^0}$, the prediction error estimate $\overrightarrow{E^p}$, and the prediction error correlations $\left\langle \varepsilon^p \varepsilon^p \right\rangle$.

If multivariate and three-dimensional data is involved, influence of all variables on any single variable and influence of one-level data on the same or other variables at other levels has to be considered. The atmospheric 3-dimensional analysis may include variables such as the u- and v-wind components, geopotential height, and height thickness (u, v, ϕ, $\Delta\phi$). The analysed geopotential height increment at a specified location is influenced by the nearby observed height increments, and the observed wind and thickness increments. The performed analysis is multivariate and three-dimensional, implying that the observed increments need not be located at the same level as the computed analysis increment. For e.g., wind increments at 250 hPa may influence the height analysis at 200 hPa. The interfacing of the variables in three-dimensional space is obtained through the specification of the appropriate correlation functions,

Because of the multivariate three-dimensional nature of the analysis, the prediction error correlations relating different variables at different horizontal and vertical locations are to be obtained. The prediction error correlation can be calculated

assuming that it can be expressed as the product of vertical and horizontal components. The horizontal prediction error correlation is calculated between the geopotential height or thickness data at one location with geopotential height or thickness data at another location. The multivariate coupling is achieved by finding the appropriate prediction error correlation functions for various combinations of ϕ, $\Delta\phi$, u, and v observations. This ensures that the geostrophic relationship between the analysis increments of geopotential height and wind is ensured, and however not in the analysed fields.
The parameters that are to be determined are the observation error estimates $\overrightarrow{E^0}$, the prediction error estimates, $\overrightarrow{E^p}$, and the prediction error correlations $\left\langle \varepsilon^p \varepsilon^p \right\rangle$. As described earlier, prediction error denotes the error value assigned at a particular location due to the errors of the forecast model, and the observation error represents both the instrument error and representative errors (measurement is simply not representative of features that can be resolved by the model). If an incremental analysis scheme is both multivariate and three-dimensional, the prediction error correlations has to be defined three-dimensionally for data increments of all different variables.

The theoretical development for the multivariate and 3-dimensssional variables is now considered. From equations (4.3) and (4.4), it can be seen that the observation increment for i is also the difference between the observation error and the prediction error at the observation location (i.e., $F_i^o - F_i^p = E_i^o - E_i^p$). Taking the product of the increments for observations i and j and averaging over a large ensemble, the covariance equation can be written as

$$C_{ij} = \left\langle (E_i^o - E_i^p)\,(E_j^o - E_j^p) \right\rangle = \left\langle E_i^o\ E_i^o - E_i^p\ E_j^o - E_i^o\ E_j^p + E_i^p\ E_j^p \right\rangle \qquad \text{.....(4.22)}$$

Because of the assumption that the observation error and prediction error are not correlated, the cross-covariance terms will be zero. Further, with the assumption that the observation errors are uncorrelated limiting to only the horizontal component of the correlation (i.e., observations i and j are at the same vertical level). So, for i not equal to j,

$$C_{ij} = \left\langle E_i^p E_j^p \right\rangle \qquad .....(4.23)$$

Anyhow, observation error correlation shall not be neglected when i = j. So, the variance of the increments, at location i is given as

$$C_{ii} = \left\langle \left(E_i^o\right)^2 \right\rangle + \left\langle \left(E_i^p\right)^2 \right\rangle = \left(\overrightarrow{E_i^o}\right)^2 + \left(\overrightarrow{E_i^p}\right)^2 \qquad .....(4.24)$$

Due to the assumption that the means of the observation errors and the prediction errors are zero, equation (4.24) indicates that the variance of the increments is equal to the sum of the squares of the observation and prediction error estimates. Assuming that these error estimates do not vary with observation location, the variance of the increments for all observations can be expressed as

$$C == \left(\overrightarrow{E^o}\right)^2 + \left(\overrightarrow{E^p}\right)^2 \qquad .....(4.25)$$

The correlation Cij or C can be estimated easily. This estimate is biased because C contains contributions from both the prediction and observation errors. The normalized prediction error correlation $\left\langle \varepsilon_i^p \varepsilon_j^p \right\rangle$ is estimated from the equation

$$\left\langle \varepsilon_i^p \varepsilon_j^p \right\rangle = \frac{\left\langle E_i^p E_j^p \right\rangle}{\left(\overrightarrow{E^p}\right)^2} \qquad .....(4.26)$$

and

$$\frac{c_{ij}}{c} = \frac{\left\langle E_i^p E_j^p \right\rangle}{\left(\overrightarrow{E^0}\right)^2 + \left(\overrightarrow{E^p}\right)^2} \qquad .....(4.27)$$

Using equations (4.26) and (4.27) we get

$$\frac{c_{ij}}{c} = \left\langle \varepsilon_i^p \, \varepsilon_j^p \right\rangle = \frac{\left(\overrightarrow{E^p}\right)^2}{\left(\overrightarrow{E^0}\right)^2 + \left(\overrightarrow{E^p}\right)^2} \quad .....(4.28)$$

The calculation of the correlation will be biased by the factor 'R', due to the term representing the square of the observation error equation (4.24), as R is given by

$$R = \frac{\left(\overrightarrow{E^p}\right)^2}{\left(\overrightarrow{E^0}\right)^2 + \left(\overrightarrow{E^p}\right)^2} \quad .....(4.29)$$

The correlation estimate Cij/C will approach R instead of 1.0 as observation separation tends to zero, which is consistent. The denominator of equation (4.29) is simply the variance of the increments.

If the value of R can be determined, $\left(\overrightarrow{E^0}\right)^2$ and $\left(\overrightarrow{E^p}\right)^2$ can be computed separately to segregate the variance of the increments as due to prediction and observation errors. From the data, the horizontal correlations c_{ij} or C for observation increments can be estimated at various distances. By plotting these curves, the value of R can be estimated for each level. The function that best fits the scatter plot of $\frac{C_{ij}}{C}$ versus distance provides the horizontal height-height correlation function.

To determine the value of R at each level, a correlation function is to be selected to fit the calculated correlations. For this purpose, several suitable functions are available. For e.g., use of the negative squared exponential (NSE) function in the applications of multi-variate optimum interpolation (MVOI) in global data assimilation (Bergman, 1979; Lorenc, 1981); ECMWF uses the summation of Bessel functions; the NMC global data assimilation system uses a third-order autoregressive (TOAR) function. A critical assessment of the statistical properties of various functions, including the NSE, second-order autoregressive (SOAR), and third-order autoregressive (TOAR) concluded that SOAR has all the statistical properties needed for an MVOI analysis correlation function which could fit to correlations estimated from observations. Further, it was also reported that the quality of the analyses produced using the SOAR were comparable to those produced using the more complicated TOAR and superior to those produced using the NSE method.

The evolution of Optimum Interpolation (OI) in the late 1950s has led to the current data assimilation method for initializing the forecast models. Data assimilation uses statistical interpolation methods that combine observations with short range forecasts to produce, as accurately as possible, a dataset that represents the current state of the atmosphere. The assimilation model produces a short forecast, usually 3 or 6 hours into the future, and then incorporates recent data into the model to adjust the model to the current observations. The concept and procedure of data assimilation concepts is explained in the next section.

4.7 Initialization

At the time when the application of atmospheric models had been shifted to the use of primitive equation models, more attention was drawn towards the problem of initialization. When primitive equations are used for numerical prediction, any imbalance between the mass and velocity fields in the initial state would reflect as anomalously large gravity-inertia waves that may persist for a long time yielding to spurious forecasts. It was noted that such imbalance in the initial fields led to totally unrealistic pressure tendency of 145 hPa/6h obtained in Richardson experiment.

Balance in the initial data is achieved by a process noted as initialization, in which the initial fields are presented in such a way that the amplitude of the gravity inertia waves remains small throughout the forecast. If the fields are not initialized, spurious oscillations will occur leading to forecast errors.

The atmospheric balance is delicate, meaning that minor perturbations may disrupt the balance, but it is also robust such that local imbalances tend to be removed through dispersion by gravity-inertia waves leading towards natural adjustment between the mass and wind fields.

Numerical forecast experiments with and without initialization have shown contrasting differences with the absence of spurious oscillations and realistic tendencies in the initialised prediction process. The initialization was achieved through static and dynamic strategies.

Hinkelmann (1959) derived the initial winds using the geostrophic relation, and integrated the primitive equations using a very short time step. Although his experimental forecasts with the primitive equations and initialization were noted to be far superior to the quasi-geostrophic model integrations, Charney (1955) proposed that the use of nonlinear balance equation may produce a better estimate of the initial wind field, as this equation includes the curvature of the streamlines implying the wind as non-divergent. The difficulty with the solution of the balance equation using the geopotential and stream function fields was the constraint that the data satisfy an ellipticity relation. To overcome this problem, Phillips (1960) suggested that an improvement of balance would result if the vertical velocity field is derived using quasi-geostrophic ω–equation and the equation of continuity. Each of these methods had rendered improvements but the noise problem persisted.

Sasaki (1958) formulated an initialization method based on variational calculus. In this method, both the wind and mass fields were constrained to fit a balance condition while remaining close to the original analysis by a suitable choice of weighting functions, with more weightage for the wind and more weightage for the height in high latitudes. Although this variational method of initialization was not widely used, this method is now the basis of modern data assimilation techniques. All the above methods are the static initialization methods.

In the approach of dynamic initialization, forecast model itself was used to define the initial fields (Miyakoda and Moyer, 1968; Nitta and Hovermale, 1969). The supposition is that the dissipative processes damp out high frequency noise as the forecast evolves. In this process, numerical integration schemes having selective damping of high frequency components can be utilised. In one of the approaches, the model is integrated forward one time-step and then backward to the initial time with the dissipation active, completing one cycle and the cycle is repeated enough times till the high frequency components were damped out. The forecast starting from this initialization is noise-free, but had the problem of damping the meteorologically significant motions as well as the gravity waves.

In another approach, the initial fields are segregated into normal mode components, and the gravity inertia waves are filtered retaining the slow moving rotational waves (Dickinson and Williamson, 1972). Although this process of 'linear normal mode initialization' would assure a noise-free forecast, it was noted that the noise reappeared as gravity waves due to nonlinear interaction of the slow waves. So the problem of noise essentially remained. Machenhauer (1977) presented the method of "'nonlinear normal mode initialization" to control the growth of gravity waves, in which the initial rate-of-change of the gravity waves were set to zero and this method remarkably helped, as the model forecasts initialised this way were very smooth and without the appearance of the spurious gravity wave oscillations. Baer (1977) developed a similar method based on more rigorous mathematical reasoning but it was more difficult to use and so the Machenhauer's method was the most popularity used.

However, there were problems to use the normal mode initialization method in the case of limited area models; wherein the normal modes at the lateral boundaries cannot be derived. Daley (1991) provided detailed discussion on the various methods of initialization such as the bounded derivative method, the implicit normal mode method and the Laplace transform method.

A new approach, the method of digital filter initialization (DFI) was introduced by Lynch and Huang (1992). This method uses an optimal filter. The method is similar to that applied in signal processing, in which low-pass, high-pass and band-pass filters are generated and applied. Essentially for the initialization problem in weather prediction, a filter is applied to preserve the low-frequency oscillations from contamination by the high frequency oscillations. This method is now being used for application in limited area models such as WRF etc.

Data Assimilation

Data assimilation is the process through which observations are blended with model forecast fields towards the generation of initial conditions for numerical model integration. In this methodology, model analysis is not produced from observations as in "objective analysis", but rather observations are used to make corrections to the model forecast which is taken as a background field. For example, the initial conditions are generated from correcting a 6-hour model forecast starting at 00 UTC using observations over a time window of 6-hours centered at 06 UTC. In this way, the short-range forecast preserves information from older observations and produce model initial conditions for the next model forecast.

In this method, a model forecast is used as background for the analysis based on the assumption that the forecast is good, coming from a good model. The model forecast would provide information between the observations, with the model fields satisfying the physical, dynamical and numerical consistencies that can be used further for a good forecast. The model forecast, thus produced, contains information from earlier observations, which permits blend of observations subject to quality control and weightage.

As of the observations, they are not analysed directly to generate the initial state, but are used to make corrections to a model forecast (assumed to be good) field. Herein, the observations are not directly subject to quality control, but the erroneous or suspect data are identified from the differences between the observations and forecast field on the presumption that the model forecast is good. In this way, a good forecast leads to good analysis and so also a bad forecast may lead to bad analysis until the flawed tendency would be corrected under the influence of considerable observations.

The differences between the observations and the forecast field, which are called corrections, are computed at every observation point. These corrections are then computed for every model grid point (meaning that they are smoothed to the model resolution) through "objective analysis" leading to the initial conditions for model integration. This implies that the weather features in the observations that cannot be resolved by the model will not be accepted into the analysis, and that the forecast corrections from dense data areas may influence the neighbouring data-void areas with improper corrections.

Considering the prime objective of producing the initial conditions that will lead to the best possible model forecast, data assimilation is intended to produce analysis that is consistent with the model numerics, dynamics, physics, and resolution. This is achieved by affecting a series of corrections based on observations to the short-range forecast. This is why the data assimilated analyses will be different for different models.

Since observations are to be assimilated into the forecast in the form of corrections, the corrected analysis is subject to the characteristics of the observations such as the density of the observations (both in the horizontal and vertical), frequency of the observations, instrument errors, directly observed fields of radial wind, radiances that are to be converted to wind vectors and temperatures, and the derived variables like cloud motion winds which are derived using algorithms.

The data assimilation procedure extracts information from the observations scattered in space and time and transform it to model grid consistent with the physics, dynamics and numerics. For this, blending of the short range model forecast with the new observations using objective analysis is adopted. In brief, new observations are checked for gross errors, and then merged with the previous forecast to produce the background analysis for the next cycle. The forecast field used as "background analysis" is also referred to as "first guess field". For this purpose, "observation increments", defined as the differences between the first guess and the observations, are first computed at the observation locations, quality control is applied on the observation increments and then objective analysis is used to derive the observation increments on the model grid. As a next step, the observation increments, as corrections, are added to the first guess to produce the new analysis. This new analysis will be used as initial conditions for the next model forecast cycle. In some of the assimilation formulations, an initialization step (such as a digital filter) will be included to remove rapid oscillations to ensure smooth run of the model. Step-wise implementation of the method is presented as follows:

Step 1: The "background field" is the short range forecast as it contains information from the model forecast based on prior observations. The background field is also called "first guess".

Step 2: Observations collected over a time period, called a time window, are used for analysis. The time window is centered on the analysis time which implies use of all observations before the data cut-off time. The characteristics of observations may vary distinctly and so are to be treated differently. Some of them may be from different instruments due to different locations; some of them may be pre-processed (such as hourly averaged winds from wind profilers), some of them may be available under certain weather conditions only (satellite radiances not available under cloud conditions and cloud motion wind vectors only in cloudy conditions), some observed variables (such as temperature) may concur with model variables whereas some (such as radiances, precipitable water) may not correspond with model variables.

Step 3: The raw data are to be corrected for known biases and grossly wrong values are to be excluded.

Step 4: Observation increments, which are the differences between the observation and the forecast value, are calculated at each of the observation location (not at the model grid points). It is supposed that the addition of the observation increment to the background field would lead to the final analysis. Since the observations are taken at locations other than model grid points and within the time window of analysis (chosen to be few hours before and after the time of analysis), the model variables are to be interpolated to the corresponding location and times of the observations. In some cases, such as satellite radiances, the model variables are used to derive the observed data variable. For e.g., model derived temperature, moisture and ozone fields are used to derive the "radiances" and these derived variables are compared with the corresponding observations. These "observation increments" are subject to errors due to instrument measurements, representativeness, interpolations and variable conversions.

Step 5: Quality control checks are performed on the observation increments and not on the observations. This would help identify improbable anomalies considering the probable variations in space and time. Some complex checks are also implemented regarding the occurrence of common errors in data, such as satellite data within the swath area having same errors (due to the observation system) in which case these observations may not be rejected (due to consistency) but they should not influence the final analysis.

Step 6: This step is the core of DA process, wherein all the observation increments related to different types of observations are determined considering the value of the observations and of the first guess field. This currently adopted objective analysis procedure is called "3D-VAR" which is a replacement of the previously used "optimum interpolation" procedure. This procedure is important as it resolves the analysis resulting from a balance between the observations and the first guess; the analysis is smooth or fluctuating; whether observations are allowed to modify the forecast variables in conformity with the model physics and dynamics and whether the observations are treated physically following trajectories or isentropes and not over a generally envisaged spherical region. This part assumes importance as it contributes to the success or failure of DA procedure.

The analysis procedure essentially weighs the balance between two evidences, one is observations and the other is the model forecast fields (first guess fields). The model error structures are determined considering the errors not only at the observation locations but also their spatial variations (different locations and different levels) and those of other variables. In this way, the observations will be assimilated into the model forecast field considering their errors sourced from instrument measurements and representativeness and their variations for different observation platforms. An example of representative-ness is the data from a radiosonde ascent within a rainband that would not be representative of the forecast model resolvable scale. Final modified analysis is created striking a balance between the first guess and the observations.

The final analysis will be determined through a comparison of the observation and background errors, as inclined to the field with smaller error (either observation or first guess). This procedure conserves the structure of the background field taking into account of the relationship between the variables and their variations horizontally and vertically. For e.g. the winds at 700 hPa may be subjected to change due to changes in the wind at upper or lower levels and by changes in the horizontal temperature gradients below due to thermal wind relationship. This is the reason why the determination of the observation increments differs with model formulation (resolution, dynamics and physics).

Step 7: The next step is to determine the analysis increments (i.e.) the corrections that are to be affected at each of the model grid point. It is important to differentiate these with the observation increments, which are the corrections identified at each of the observation locations. These analysis increments are produced considering all of the observations through the objective analysis (3D-VAR) method. Since the model forecasts are supposed to be good, these are considered as "corrections" to the first guess field. These analysis increments are smoothed enough to exclude variations associated with scales smaller than that the model can resolve, thus identifying the areas where the corrections are to be adopted and not.

Step 8: This step is to determine the final analysis that would serve as the initial conditions for the model's next forecast cycle. This analysis thus retains all of the information contained in the first guess that is compatible with the observations. Large gaps in the observations may lead to exclusion of important and retention of wrongly predicted small scale features in the analysis, and a priori knowledge of model predictability would help understanding of how the observations would influence the analysis.

Consideration of the differences of the observations and analysis (which denote that part of the magnitude not taken into the analysis), smaller differences with random pattern indicate acceptance of all consistent observations into the analysis. Whereas, large differences, indicative of representative error, even though are assimilated would not be retained in the subsequent model forecasts due to model resolution.

Forecast model: The successful implementation of the DA system depends on the forecast model, as the model constitutes an essential component. Essentially, the forecast model produces the atmospheric structure in four dimensions in consistency with the dynamics, physics and numerical formulation. The model helps fill the data gaps as it gets integrated forward from previous observations to future time subject to the limitations of the model that would affect the prediction.

Cycling: The model forecast system is repeated in cycles, so as to carry forward the variable information from the past to the present conforming to the model consistency. In this process of "cycling", the model forecast and the data assimilation procedure march together forward. The cycling process may sometimes retain the errors due to bad observations for several cycles before they are removed or good data are ingested. For the same reason, the model retains any of the high-resolution model generated spurious small scale features as they do not get corrected if the observations do not have sufficient resolution to correct them. The cycling process also helps with the spin up problem which arises when a new forecast is initiated. The cycling process is also essential to preserve balances of the variables from the physical parameterizations. For e.g., if the relative humidity field has imbalances with model generated cloud distributions, meaning if the relative humidity values are smaller (higher) than the model thresholds for cloud existence where the model predicts (does not predict) a cloud, the model tends to evaporate the clouds and reduce precipitation or create excess clouds and increase precipitation. The analysis thus produced will have enormous impact on the quality of the forecast, as it will be the basis of the initial state for the forecast run. As noted earlier, the short range forecast will be the first guess for the DA procedure. Generally it is not and instead, the initial state of the forecast cycle is used for carrying out a separate analysis and forecast run to include more observations that would become available after the cut-off time so as to make the short range forecast better and thus producing a better first guess field for the next operational forecast cycle.

The importance of the forecast model in a DA system can be understood as the model should be capable of porting all important atmospheric information from the previous observations along with producing good prediction of important atmospheric circulation patterns such as fronts, lows, convection, boundary layer etc. Sometimes this is achieved through carrying out the assimilation procedure in short time scales successively to assimilate all available observations in shorter time windows. For e.g., one hour assimilation cycles could be carried out three times and the analysis at the end could be used as the analysis (initial state) for the next operational short range forecast (for the next 24-hours).

Data: The objective of the DA system is to modify the first guess field, from the short-term model forecast, through blend of observations. Several types of observation data from different platforms and conditions are used by the DA system. For e.g., radiosonde data is taken at 12 hour interval and over land; visible satellite cloud drift data during daylight hours and only over cloudy areas and data only at cloud tops; polar orbiter data twice a day over tropics and middle latitudes and over the swath; satellite radiances data only over sea; IR satellite radiances only in clear weather conditions. The limitations on the observations during bad weather conditions may lead to gaps in data over regions of important weather systems. For e.g., data void regions due to the presence of a deep synoptic system that lasts for a few days.

The data of observations, received in coded form, are to be deciphered and subjected to checks to eliminate errors due to instruments, communication and transcription. Decoders are used to convert the data to convert all data into a common format such as BUFR ((Binary Universal Form for the Representation of meteorological data) format set by WMO. The raw observations are checked for instrument-errors before they get into the data stream. Then the data are checked for gross errors (physically impossible values) and duplications. Similarly biases in observations are also identified, such as biases in humidity that differ with manufacturing agency; biases in ACARS (Aircraft Communications Addressing and

Reporting System) data. Even after the checks for errors and biases, some of the observations may still have small to moderate errors which may not be compatible with other observations and/ or first guess field. These data are subject to quality control checks of the DA system and also manually.

The data that will be used have advantages and also limitations. Properties of some of the data may help understand their usefulness. Multi-level mass and wind data from RAOB and ACARS (ascent and descent profiles) allow the most consistent changes to the background fields and ageostrophic features, which are well retained in the model forecast. Surface observations also have limitations. Surface observation locations may be at different heights, not coincident with model topography, which may impact all observations, especially surface pressure. There may be problems to derive vertical structure, for e.g., where the observation shows lower than predicted value. This necessitates assumptions for relating the surface values to lowest model layer (such as PBL). These are some issues in effecting the corrections to model background fields using surface data. Satellite based data, such as radiances, represents radiation from a deep layer of the atmosphere, and so with coarser vertical resolution than the model. In this case, deficiencies to properly vertically distribute the deep layer average information can impact the model forecast. Proper assimilation of satellite radiances needs accurate information on the emissivity which may have problems over land as the present methods are reliable for channels sensing the upper troposphere and stratosphere. Satellite winds are given at a location only and not the vertical structure. Assignments of heights for these observations may have errors and the error may be large in regions of strong shear. Cloud and precipitation data from satellites and radar will be used to correct the model forecasts to add or remove them. Precipitation observations will be used to correct model soil moisture. However, limitations exist with regards to the correction of model humidity fields as how much at what levels, how much cloud water should be added or removed and with what distribution, correction of vertical motions as consistent to diabatic heating. As such use of clouds and precipitation information can only be affected concurring with good information on heating and associated vertical motions and convergence/divergence patterns. Otherwise, this procedure may lead to imbalances and erroneous forecast. This correction is more beneficial during the early part of forecast (first 12 hours). It is also possible to have problems with resolution higher than the mode (i.e.) higher density (such as from satellite and radar). In these cases, errors in observations may dominate to affect the analysis. To overcome this problem, it is suggested that the density may be reduced by choosing only a limited observations (average of nearby observations) which are representative and compatible with model resolution. In order to increase the utilisation of high density observations, the analysis should be designed to permit small scale changes in the first guess field. Otherwise, with border background covariances, high resolution observations may not be able to affect the analysis so as to represent small-scale phenomena.

Quality control: Firstly, observations are rejected if they are sourced from locations or equipment which are previously identified to have problems with observing techniques; and/ or if the difference between station height and model topography is large. The quality control process of the DA system has "Buddy checks" and "Complex Quality Control" to ensure consistency between variables and in the vertical profile of a variable. It is to be noted that an observation is excluded if the disparity between observation the analysis is large, but the same would be reconsidered if it fits better into a modified analysis during the DA iterations.

As per the description, the analysis is carried out as a series of iterations. With each of the iteration, the background field will be modified towards a better guess all subject to the cost function. In this process, intelligent quality control means not to reject a doubtful observation (due to large observation error). In this way, if other observations tend to modify the analysis favourable to the observation increment that is doubtful, the previously reject observation may now fit into the analysis. This will ensure that good observations which are substantiated by other observations will mostly be absorbed into the analysis. This nonlinear quality control method is now being adopted in NCEP GFS analysis.

Observation increment: The forecast fields are used to derive the values at the observation locations, and the differences between the observations and the model based values at each of the observation location are computed, which are called "observation increment". These observation increments are subject to quality control (not the observations themselves). Although this looks simple and straight forward, it is to be noted that some of the observed data may not be representative because of their sampling pattern. For e.g., a radiosonde ascent giving data at few seconds interval may be providing samples within and out of the cloud as it drifts upward in a thunderstorm environment; wind profiler data containing gravity waves and the mean flow; and satellite radiances representative of wide space and deep layer but at time scale of seconds. In contrast, model forecast data is sampled at the model time step and representative of the 3-dimensional space of the model grid. The model fields are to be interpolated to the observations at the different locations and times. The interpolation may sometimes lead to considerable disparities which either affects as a large correction or a complete rejection. The method of interpolation also affects the model based value, such as linear interpolation affects amplitude and higher order interpolation may generate unrealistic maxima or minima. Since all the observations contained in the time window are used in this process, the computed observation increments falling within the time window may also lead to disparities due to sharp spatial variations of some of the variables in time. Frequent updated cycling, meaning that new analysis will be obtained at frequent time intervals such as 1-hour instead of 3 or 6-hour cycle, will help reduction of data rejection and enhances model consistency. At the same time, it should be noted that these methods may not ensure that the observations and analysis may not represent the same phenomena.

In the case of the variables where the observations do not correspond with the model produced variables directly, the method of comparison is to be ascertained appropriately. In the case of observations such as satellite radiances, the model variables of temperature, humidity and ozone are used to calculate radiances which are then compared with corresponding satellite radiances. In this way, upwelling radiation will be computed for different wavelengths which facilitate direct comparison with observed radiances at respective wavelengths. In contrast, conversion of satellite radiances to vertical soundings of temperature will be subject to errors from the approximations in radiative transfer physics, blend of radiances from different channels, and averaging to model levels. However, in some instances, there may be some other problems such as transforming model conditions near the surface wherein large differences in topography of observation location and model vertical levels.

The next step is to affect quality control on the observation increments, with the background knowledge that the model fields are used to generate values at the observation locations through appropriate space and time interpolations. Since the observation increments reflect the disagreement of the model fields with the observations which may be called "corrections", and knowing that these corrections could be due to observation errors resulting from instrument and representative errors and/or model forecast errors due to coarse resolution and interpolation.

In the data assimilation process, observation increments are assessed for quality control and not the observations. The reason being that the observations may show strong variations in the horizontal (such as in a trough) but the corresponding increments will be uniform if the model predicts the pattern correctly; only one or two observations may be reflecting the correct value in agreement with the model forecast but will be rejected whereas the corresponding observation increment will not be rejected. A little more detailed consideration shows that (i) in the case where both the observations and the model prediction are good, the regions of small and random variations in the observation increments need minor corrections and regions of organised patterns need significant corrections; (ii) if the background is bad and the observations are good, the observation increments will show systematic pattern; and (iii) if the background is good and one or two observations are bad, observation increments will be small all over except at the bad observation locations. These are the reasons why observation increments are considered for quality control in the place of observations.

In addition complex quality control checks are also performed. For e.g., satellite or aircraft observations taken over a large area with instrument error leads to large differences in the model forecast and these observations. Buddy checks considering other types of observations such as radiosonde or other aircraft or other systems would help identify these errors. Similarly, several methods are included for detecting coding errors and missing observations. Further, nonlinear quality control methods are adopted in which a rejected observation is re-evaluated during the iterative process of DA procedure to check if the observation could be acceptable as the neighbouring values (horizontally and vertically) are updated.

It may be inferred that, observations which are not compatible with the model forecast due to differences in the resolutions (observations within a thunderstorm and model resolution cannot include them) are to be excluded. Similarly differences may arise due to time interpolation of model fields during which period there is a rapidly evolving weather system; or the time window is long enough affecting the representation of small and fast moving weather systems; and presence of few data points in data sparse regions prohibit validation with other observations.

Finally, it would be desirable for the forecaster to assess the inconsistencies in the observations by identifying (i) data sparse regions, (ii) assessment of the earlier prediction over the data sparse area, (iii) consistency of observations between locations and other systems (such as satellites) and (iv) identify developing systems if any or features smaller than those the model can predict. This part would be more effective as the procedures improve to include provision of detailed information on the rejected data.

Analysis: The next step is to use the "observation increments" to derive the analysis. This is the most important part of the data assimilation procedure, which is also complex, as it would denote the quality of the model forecast. The objective analysis procedure derives the model grid values at the initial time by merging the observation increments computed at different observation location at different time within the time window. In the objective analysis method, each of the observation is given a weightage depending on its distance from the model grid (such that nearer values will have more weightage). In the present data assimilation method, the "observation increments" are interpolated and corrections are made to the previous short range forecast values to produce the new initial conditions for the next forecast cycle. This analysis is made so as to retain as much of the observations' information along with retaining as much of the structure of the model forecast fields.

Assignment of weights is achieved through optimization striking a balance between differences in the analysis and observations and changes in the background guess field. The basic premise of the 3D-VAR method is that it knows the behaviour of the errors in the observations of each variable and also the model behaviour in terms of the model error statistics. The balancing point is achieved through an iterative procedure of assigning penalties to the O-A (observations minus analysis) and the F-A (model forecast minus analysis). Penalties will be evaluated considering the differences of O-A (higher the penalty if a priori the observation is known to have smaller error), and similarly penalty is assessed for background field deviations from the background field (a bigger penalty if the forecast is usually good). After several iterations, this procedure will lead to an analysis that will have the smallest combined observation and background field penalties. This will ensure that the final analysis will imbibe all better quality data and retain most of the background information conforming to dynamical consistencies. It is also to be noted that the final analysis will be deficient if the error statistics are not representative of the in situ weather.

The standard 3D-VAR method will ensure that the observations tend to influence the analysis in both the horizontal and vertical similarly, in all the weather situations irrespective of any severe weather system such as fronts. Some of the other 3D-VAR methods used at some of the operational centres are tuned to produce more adoptable analysis.

It is also important as an observation increment influences the analysis, which is dependent on the background error covariance that denotes the relationship between the errors in the first guess field. The standard 3D-VAR method assumes a circular pattern of the background error covariances, and so corrections are applied in a circular pattern (for e.g., cooling the guess field uniformly in all directions with weightage based on distance). This type of correction may sometimes be detrimental as it corrects in one region and spoils over another region. A better method would be to use "anisotropic background error covariances" and apply the corrections appropriately, such as by stretching along isentropes or streamlines.

The objective of the data assimilation process is to find a balance between the observations and the background field towards producing an optimized estimate of the atmospheric state through iterative solution of a cost-function given by

$$J(x) = J^o + J^b = \frac{1}{2}(y - y^o)^T (E + F)^{-1}(y - y^o) + \frac{1}{2}(x - x^b)^T (x - x^b)$$

The 3DVAR problem is summarized as the iterative solution of the above equation to find the analysis state that minimizes $J(x)$. This solution represents the nearest possible estimate of the true state of the atmosphere with minimum variance from the background (previous forecast) x^b and observations y^o (Lorenc 1986). The fit to individual data points is weighted by B, E and F which are the background, observation (instrumental) and representivity error covariance matrices respectively.

In simpler form, for brevity of easy understanding, the above equation could be expressed as

Final Analysis = $\Sigma(O - A)^2 + \Sigma(F - A)^2$

The first term on the right hand side denotes fit to observations and the second term works to preserve model consistency.

The following stepwise procedure is carried out to achieve the objective.

Adjustment of background fields to observations. Firstly, the analysis is equal to background (short-range forecast). This means that the 2nd term is zero as A = F. Difference between the observation and analysis denote the penalties and so the analysis is forced towards observation. This will tilt the balance toward the observation as analysis.

Now that the term (2) is large after step 1, the analysis should be moved towards the background field. The first part of moving back toward the background field is restoring its vertical structure. Restoring the background structure is critical for maintaining consistency, which is accomplished by making similar corrections to the background over a range of levels rather than just at the observation level. This spreads the influence of the observation. The vertical distributions of the model error statistics will be used to apply the corrections at different levels. This happens through some mathematical manipulations. The term (2) will be too big because the typically good forecast has a small expected error, and term (1) will be zero because the analysis has been corrected to the observation value. Therefore, the balance needs to move more toward the background and away from the observation.

Now the balance information from the observation will move towards the background. The analysis profile will change the value to lie between the background and the observation. This correction will depend on the errors of the observations and the model fields. The iterations move back and forth as corrections in terms 1 and 2 till a balance is achieved. The balance point is achieved where the "cost function", which is dependent on total penalties for not fitting the observation and the background, is minimised. It should be noted that although observation at one point is considered, corresponding variable values at many model levels that would influence the observation need to be considered for correction. As such the analysis profile would move more toward the observation.

The analysis includes adherence to dynamical consistencies such as mass-wind fields. For e.g., changes in wind analysis require changes in the temperature field. Since this manifests in the cost functions of other fields also, the other fields (such as temperature) are also to be changed. This is complex, but for mass and wind field adjustments, thermal wind relation may be used.

So steps (i), (ii) and (iii) are repeated for a second observation, for e.g., temperature at a lower level to conform to step (iv). The correction of one temperature observation at one level will modify the temperature profile as of the wind described above. Warming at one level will again require corrections to wind analysis. These wind changes affect the wind profile at the first observation location.

After these series of adjustments, the analysis will be compatible with the observations and the dynamical consistencies. To summarise, the wind observation at the first location led to wind corrections at that location and temperature corrections at a second location. These temperature changes in the temperature at the second location will affect changes in the background field of temperature at that location as well as different changes in the background field of wind at the first location. In this way, both the wind and temperature will change their background fields. The balancing act is achieved by applying the cost function simultaneously for all observations and background fields retaining maximum useful information from the observations and background fields subjecting to dynamical consistencies so that model would retain the new information.

The final step is to further consider the inter-relation of variables considering the effects of changing the assumed observation or first guess errors.

As the wind profile changes, so does the linked temperature profile. A closer fit to the wind observations will lead to a closer fit to the temperature observation. Likewise when the temperature field changes, the linked wind profile changes. A closer fit to the temperature observation will lead to closer fit to the wind observations.

Good independent observations of the same weather system normally bring the analysis closer to the observations, bringing out as much usable information as possible from the observations and at the same time not losing valuable model information.

Assumptions: In the data assimilation process, the following adopted assumptions will have impact on the analysis.

Statistical assumptions: The balancing between the observation and forecast errors are based on statistical distributions. For e.g., every time a radiosonde is launched, it may have instrument error and the forecasts also vary from day to day. A plot of these would yield a distribution. The analysis will be affected by the error distributions assuming Gaussian distribution. These assumptions influence how the analysis is modified for different data types, their horizontal and vertical spread and inter-relation of different variables. Generally these assumptions are good for common weather variations but have deficiencies for extreme weather events.

Observed error: There are two sources (i.e.) instrument error and representative error. The instrument errors are generally known and quantified. Representative errors are not well identified, as they are dependent on model resolution. If the model can resolve the weather systems that contribute to the observations, these errors would be small. This means the representative errors vary with weather situations (which depend on the amplitude and scale) and so indicate desirability of finding these errors with respect to each type of system, which is not possible. The analysis considers both of these errors to be combined and treated as one number.

Background error: The background error is dependent on model performance statistics. If the model performance is different than the in situ situation, the common BE may not be appropriate. The model error statistics may be averaged, in which case these may not be representative of regional variations. The background error consists of two parts, one corresponds to a single point and the other representing the correlations between different points (called covariances). This implies that if a model error at one point is well correlated with another point, a correction at the first point indicates a needed correction at the second location. Thus the Background Error covariance will spread information from the observations uniformly in all directions which may not be appropriate in some instances.

Balance constraint: If the analysis changes either the model wind field or the mass field only, the balance between these two fields will be disturbed. Then the model dynamics will drive the model fields toward a new distribution of mass and winds which may imply loss of information in the analysis that may constrain the forecast. This means that analysis changes to the wind field should be accompanied by analysis changes to the mass field and vice versa. The mass-wind balance may be achieved using simple or complex dynamics. Herein, the mass-wind balance is not strictly enforced which adds penalties to the cost function. An initialization procedure can also be used

to generate balanced initial conditions for the model forecast, as important information may be lost if balance is not enforced.
Tuning: The analysis is subjected to tuning to produce optimised forecasts. Tuning means changing the relative importance of different components, such as fit of temperature (or moisture) to the background, changing the distance of influence. This may affect the analysis including the dynamical consistencies and balance, which could influence forecast accuracy and the model errors. Tuning would appear as multipliers in the cost function equation, thus affecting changes in the influence of all observations compared to the background, the balance constraints and the horizontal and vertical extent of influence. The changes due to tuning may vary by analysis variable and would not affect the limitations of the analysis process and the data sources of the variables.

Summary

In the data assimilation procedure, the model background field (first guess/short-range forecast) is compared with observations at different locations and times.

The objective analysis procedure (such as 3D-VAR) will produce a smoothed field on the model grid domain using information of all the observations of different kinds and with different error characteristics, conserving the dynamical, numerical and physical consistencies in the modified background field.

The corrected analysis will be compatible with the observations to the extent that the observation errors permit. This means that the analysis gets corrected depending on the error characteristics of each observation type, with higher weightage to lesser-error type observations than those with known higher errors. The correction method will be more effective where the data is uniform. Single point data (such as radiosonde) and layer-averaged data (such as satellite radiances and precipitable water) are handled differently.

The analysis effects corrections, to the limit of the forecast errors only, to the first guess field. The modifications of the first guess field depend on the different forecast error patterns in the flow regime. A lot of good observation information that is persisting is needed to correct a bad prediction in the model first guess field (short range forecast). The analysis process ties up corrections in the wind field and temperature at a location affect corrections in their fields at other locations in horizontal and vertical in order to keep dynamical consistencies. In this analysis procedure, observations make corrections to the short-range forecast field continually. The model is assumed to generate good forecasts, as the analysis will be bad if the forecast is bad. The quality of the analysis is also dependent on the predictability of the variable, (for e.g., better predicted temperature, wind and pressure and badly predicted moisture etc.). The analysis procedure may not fit all the observations to the full extent, as the analysis should lead to the best possible model forecast. Observations will be used to correct the model resolvable scales of motion only. The assumptions of the analysis procedure are effective under common conditions and fail for extreme weather situations (rapidly developing systems).

4D-VAR: is an extension of 3D-VAR in time. In the 3D-VAR, the first guess field is used to derive the variable values at the observation location and time by executing interpolation in space and between forecast times. The "observation increments" (differences between the first guess values and the observations) thus derived correspond to one and the same time. In 4D-VAR the "observed increments" are allowed to evolve following the model dynamics and physics from the observation time to the analysis time. In the 4D-VAR, modification of one variable affects all other model variables. In the 3D-VAR, the background error covariance will not change, whereas in 4D-VAR, flow-dependent background error covariances will evolve which are consistent with the model dynamics and physics for the in situ weather conditions. Thus the analysis increments in 4D-VAR are expected to be more structurally detailed, better coupled among all model variables, and make better use of more observations from the same sites over time.

One of the limitations of 4D-VAR is that it requires large computer resources and so it is generally run at coarser resolution than of the in situ model. Another limitation is that any of the deficiencies in model physics will reflect on the evolving background error covariances that affect the analysis leading to errors in the forecast. It is difficult to identify these errors due to complexities in model physics. ECMWF has been using 4D-VAR since 1997 but has always been running it at coarser resolution than their forecast.

Ensemble Kalman Filter: This is more efficient and robust as it can improve the covariances as part of the analysis itself depending on the observations. However the Kalman filter method is suitable for linear systems and the considered atmospheric prediction is a nonlinear system. This method also requires large computer power. As a simplification, the covariances are statistically determined using ensemble forecasts following the principle that the forecast error is proportional to ensemble spread. This makes the first guess error field to be similar to the ensemble spread. The Ensemble Kalman filter (EnKF) method provides means to generate perturbations for generating the ensembles and the production of ensemble forecasts as required. A limitation of EnKF is that fewer ensemble members will restrict the degrees of freedom in the generation of background error covariances. From the methodology, it is to be expected that this method will not be good where the model deficiencies result in large errors with small spread. The small spread will lead to less weightage to the observation increments, thus do not help to correct the first guess field appropriately.

5. Parameterization of Physical Processes

NWP models use the governing equations of atmospheric dynamics and physics written in mathematical form and solved using numerical methods. Numerical weather prediction models have three sources of error: (i) truncation errors due to approximation of differential equations by finite difference approximations or spectral representation; (ii) errors in initial conditions of the atmospheric models; and (iii) errors due to approximations of physical processes such as radiation, convection, PBL etc. and the contributions due to various errors were reported as 48%, 18% and 34% respectively. As such all the scales of atmospheric motion cannot be resolved by the models due to constraints from truncation. These atmospheric models cannot resolve weather processes that occur within a single model grid box.

A global infrared image from a geostationary satellite (figure 5.1) shows features of the atmospheric phenomena with wide range of horizontal scales, such as extensive convective activities over the tropics suggested by a cloud distribution. These convective activities involve a wide spectrum range. The smallest scale of those corresponds to individual cumulus convective towers (or hot towers), and the individual convective elements are organized into the mesoscale, and these mesoscale organizations are further organized into a planetary scale representative of Madden-Julian oscillation. As a whole, the image suggests that a wide spectrum of convective cloud variability involves intensive interactions crossing the scales. These individual convective towers (hot towers) are hardly resolved by a typical model. As a result, contribution of those convective towers to the global tropical dynamics must somehow be accounted for in indirect parametric manner. This problem is called the parameterization.

Figure 5.1 Infrared image from a Geostationary satellite (from COMET)

Parameterization schemes are designed using observations and the laws of physics. For e.g., model representation of cloud microphysics that is governed by the conservation of the mass of water in the volume of a grid box. Water vapor in the model condenses according to observationally based thresholds of relative humidity, forming "clouds". Physical processes lead from clouds to the formation of droplets that then fall to grid boxes below as rain or snow, also based on observational evidence from within clouds. Thus, the parameterization of clouds, although subject to many unknowns, is based on a combination of the conservation of mass and energy and an empirical understanding of cloud formation, based on observations. Using those properties, modellers create parameterizations for cloud processes that operate within model grid layers when appropriate environmental conditions are satisfied.

It is true that parameterizations add to model prediction uncertainty. Some of the physical processes are better understood than others, but this is true of all model processes. Even though it is well known that uncertainties due to numerical representation of the laws of motion does exist, understanding of these uncertainties in weather models is decreasing due to increased understanding of weather processes.

Even in high resolution models, some of the weather processes may be of a smaller scale that cannot be calculated explicitly even though they influence the weather systems. These sub-grid scale processes, such as radiative transfer processes of shortwave and longwave radiation, cumulus convection, vertical energy fluxes in the planetary boundary layer etc. indirectly affect the resolvable scale weather variables.

Since the basic idea of parameterization is to represent the subgrid-scale processes in parametric manner, it suggests simplification of the physics so that the physics is not described explicitly, but remains implicit in the formulation. Thus, any physical process that cannot be predicted explicitly necessitates a parameterization scheme based on justifiable physical (for e.g., radiation) or statistical (for e.g., inferring convective cloud fraction from convective precipitation amount) formulations. The parameterization scheme should derive information about these unresolvable scale processes from the resolvable scale variables in the prediction system using appropriate assumptions. Generally, the word "closure" is used to define the assumptions in the parameterization that relate the subgrid-scale variables with the resolvable grid-scale forecast variables. (It closes the loop between the parameterization and forecast equations). In fact, several assumptions may have to be used to define the "closure". Three forms of the assumptions are stated as

5.1 Empirical/Statistical

These relationships are assumed to be true for every case (for e.g, variations of surface layer wind speed with height for PBL processes as needed for surface wind forecasts). It may be noted that for a normal statistical distribution, expected outliers could be 5%.

5.1.1 Dynamical or Thermodynamical Assumption

A complex process that is expressed through a simplified relationship, for e.g., quasi-equilibrium assumption in Arakawa-Schubert convective parameterization.

5.1.2 Model within a Model

Even if nested models are used, with the derivation of information from coarse to fine resolution, assumptions are still to be applied at different resolutions.

The main issue of parameterization is the attempt to predict (with insufficient information), the effects of subgrid-scale processes using the information at the grid scale, for e.g., prediction of wind in a grid box to predict the boundary-layer turbulence without the details of topography, vegetation characteristics, or other details at the surface. Some of the concerns with using parameterizations may result from:

Interactions between the parameterization schemes, where each scheme will have its own errors and assumptions (for e.g., a soil model and a radiation scheme exchanging information of heating the boundary layer)

The forecast errors, which arise due to complexity of interrelated parameterizations, are also difficult to trace back. The largest impact of using parameterization schemes generally reflects on the predictions of sensible weather at the surface. These issues are to be carefully considered wherein the parameterized physical processes are important for the forecast. As the modellers tend to include more of the atmospheric processes to be parameterized with an intention to add more details, it is to be remembered that the forecast will also be more sensitive to the physical parameters whose values are not well known.

Adding more of the parameterizations will also take longer time of prediction due to complex interactions between different parameterizations. It may be true that the model skill will improve and model predicted phenomena will be more realistically realised, more sophisticated schemes and finer resolution may lead to more realistic-looking forecast but at the same time the model error characteristics will also be more complicated. As model error characteristics become complex, adjustment of model predicted fields will be more dependent on model diagnostics.

Even in high resolution models, some of the weather processes may be of a smaller scale that cannot be calculated explicitly even though they influence the weather systems. These sub-grid scale processes, such as radiative transfer processes of shortwave and longwave radiation, cumulus convection, vertical energy fluxes in the planetary boundary layer etc. indirectly affect the resolvable scale weather variables.

It is true that parameterizations add to model prediction uncertainty. Some of the physical processes are better understood than others, but this is true of all model processes. Even though it is well known that uncertainties due to numerical representation of the laws of motion does exist, understanding of these uncertainties in weather models is decreasing due to increased understanding of weather processes.

A substantial amount of the current research on parameterization is devoted to improving the conceptual models and closure assumptions to correspond with increasing the resolution and numerical treatments resulting from increasing computer power. So, synchronous advancements in the representation of physical and dynamical processes and numerical methods along with increasing computing power are important for numerical weather prediction.

5.2 Atmospheric Radiation

Solar radiation is the main source of energy of the earth-atmosphere system. Incoming solar radiation (shortwave) and the longwave radiation from the atmosphere contributes to heating of the earth's surface and outgoing longwave radiation from the earth's surface contributes to cooling and a balance of these two processes determine earth's surface temperature. So, atmospheric models should incorporate these processes with well-defined physics.

The processes of solar (shortwave) and terrestrial (longwave) radiation should include the interference of clouds, water vapour, trace gases, and aerosols. Similarly, the surface characteristics of earth (both land and sea) such as vegetation type, soil type, soil moisture quantities, snow, water bodies, and land and sea ice etc. and their interference with both the solar and terrestrial radiation are to be included.

Since all the radiation processes cannot be explicitly included, as solar radiation is spread over a spectrum of wavelengths and its interference with the earth's atmosphere subject to absorption, reflection and scattering; as terrestrial radiation depending on the land surface which has several types; and the absorbing characteristics of radiation by the atmosphere being different for the various atmospheric constituents, these processes are to be parameterized.

The importance of radiation processes in weather prediction is dependent on the meteorological situation, as it dominates the advection process for e.g., in a high pressure system or clear sky night (advection forcing dominates such as in developing pressure systems). The radiation processes are also significant when the diurnal cycle of solar radiation is large (as in summer) near the ground where the parcels are subject to sensible heating and where precipitations processes release latent heat (or evaporative cooling).

Although the processes of atmospheric radiation are well known at small time and spatial scales, difficulties arise when the variations are to be forwarded to coarser model resolution. While the amount of incoming solar energy at the top of atmosphere and its variations on diurnal, seasonal and climate scales are known, uncertainties exist with regards to its absorption, transmission, reflection, scattering in the atmosphere and of absorption (by different surfaces) and emission of longwave radiation at the earth surface. These uncertainties are due to difficulties in defining the effects of atmospheric constituents on the incoming solar and outgoing longwave radiation; due to difficulties in defining energy fluxes and defining surface characteristics such as soil moisture at the earth surface. For e.g., systematic model biases such as prediction of higher precipitation lead to higher soil moisture which again causes increase of model precipitation. Similarly improper description of the surface characteristics also lead to errors (for e.g., a grid box containing 2 or 3 types of surfaces such as an urban area within a forested area may be treated as deciduous forest).

The characteristics of the shortwave and longwave radiation are well understood, as they take place on very small time and spatial scales and are influenced by the atmospheric composition. Solar radiation energy comes in ultraviolet, visible and near-infrared frequencies, with peak at visible wavelengths. As the solar radiation passes through the atmosphere, it is subjected to absorption, reflection and scattering. Some of the reflected and scattered radiation also reaches the ground as diffuse radiation. It may be considered that about 50% of the insolation (incoming solar radiation) at the top of the atmosphere reaches the ground. Radiation energy is emitted back to space from the earth surface at longer wavelengths, and is subjected to absorption by the greenhouse gases, clouds and aerosols. This absorbed energy in the atmosphere is reemitted in all directions. Because of this process, earth's surface temperature is noted to be higher by ~33 C.

The following figure shows the energy spectrum of the solar radiation at the top of the atmosphere and at the earth's surface under general conditions.

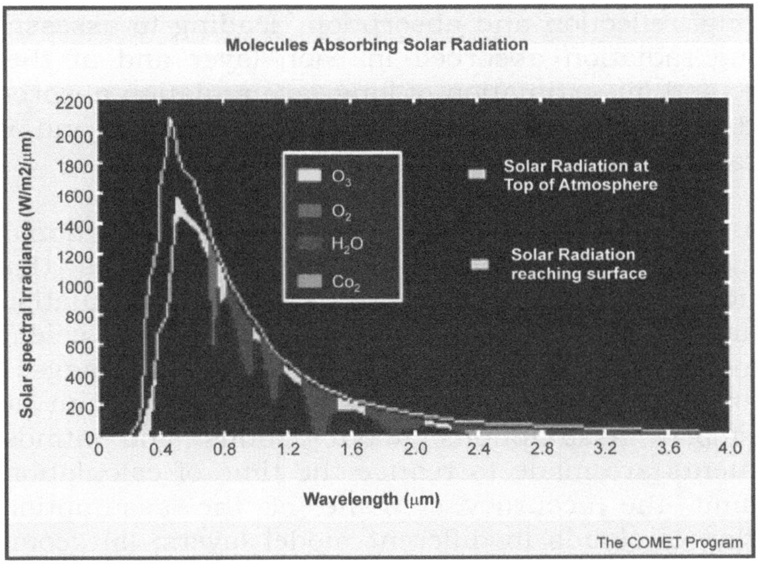

Figure 5.2 Energy spectrurm of the solar radiation (from COMET)

The incoming shortwave solar radiation is absorbed by ozone, oxygen, water vapour, and carbon dioxide. The absorption of shortwave radiation takes place at certain wavelengths, and the amount of absorption depends on the total amount of the absorbing gas present in the direct path between the top of the atmosphere and the surface. As could be seen, some of the radiation is subject to scattering and reflection by air molecules, aerosols, and clouds. The absorbed radiation is reemitted as longwave radiation.

The longwave radiation emitted from the earth (and the small amount of near-infrared solar radiation) is also affected to absorption in the area of the longwave spectrum, in the region of 3.7 and 16.7 microns. The absorbed longwave radiation is reemitted both upward and downward, but at a different wavelength and intensity based on the temperature. Since the actual absorptions and reemissions take place on molecular scales, which cannot be included precisely, these processes are to be parameterized.

Parameterization of the radiation processes in atmospheric models requires (i) division of the atmosphere into vertical layers and determining the amounts of cloud, absorbing gases and aerosols in each layer; (ii) estimating the effect of scattering, reflection and absorption leading to assessment of incoming radiation absorbed in each layer and at the earth surface; and (iii) estimation of longwave radiation absorbed and emitted by each layer (considering the amount and in situ temperature and pressure).

As the radiation calculations take large computational time, they are carried out at much lesser intervals than the dynamical calculations (for e.g., radiation calculations are performed at 1-hour interval whereas dynamical calculations are done at every 1-minute). For this reason, changes in the solar angle are included in the radiation calculation at every time step. Assumptions about clouds and atmospheric constituents are made to reduce the time of calculations, but they limit the accuracy. Some of the assumptions are (i) overlap of clouds in different model layers; (ii) geometry of clouds, cloud particles and aerosols; (iii) water phase (liquid or ice) in cloud layers; (iv) relationship between relative humidity and cloud water mixing ratio to fractional cloud area; and

(v) relationship between amount of precipitation and fraction of grid. Due to these assumptions, large errors occur from the estimation of clouds and cloud fraction which will further lead to errors in radiative transfer calculations.

The remaining part of the insolation, after scattering, reflection and absorption in the atmosphere, will be subject to reflection (surface albedo) at the earth surface. The remaining part of insolation and the longwave radiation emitted from the earth surface, both of them dependent on the differing surface characteristics of surface type, surface roughness and soil moisture will contribute to net energy balance and the surface temperature.

The surface energy balance will have the components of (i) net insolation absorbed at the earth surface; (ii) net longwave radiation resulting from emission from the earth surface and absorption of longwave radiation from the atmosphere at the earth surface; (iii) amount of energy utilised for heating the surface and sub surface; and (iv) sensible and latent heat flux from the ground.

In the atmosphere, earth's surface interacts with incoming solar radiation that remains after scattering, reflection, and absorption by the atmosphere. The resulting surface energy balance depends upon surface albedo, availability of water to evaporate from the surface and/or its vegetation, roughness of the surface, surface type (soil, water, or ice), presence of snow, and other characteristics. The net surface energy balance directly determines surface temperatures and characteristics of the atmospheric layer directly influenced by the earth's surface (the planetary boundary layer or PBL). All of these components are dependent on the surface characteristics such as type (land, water or ice), type and amount of vegetation (needed for albedo, evapotranspiration), type of soil and surface roughness. The energy and water balance at the surface are linked through evaporation.

In atmospheric models, parameterization of the surface processes will consider the interaction of incoming solar radiation with earth surface to derive the exchange of heat, moisture and momentum between the earth surface and atmosphere which also drive the planetary boundary layer. For the surface parameterization implementation, it is important to know of the land surface scheme adopted, sources of surface characteristics and their interaction with the atmosphere, the number of soil layers etc. as they are used to derive the soil moisture and temperature and errors in these process representations lead to errors in the computation of surface temperature and moisture.

The passage of radiation from the earth surface to the free atmosphere happens through its transport across the planetary boundary layer and so the processes for transport of heat, moisture and momentum in PBL are important for radiation estimation. The parameterization of PBL is separately described.

5.3 Atmospheric Boundary Layer (ABL)

The Atmospheric Boundary Layer (ABL) or the Planetary Boundary Layer (PBL) is generally defined as the lowest portion of the atmosphere that is in contact with the earth's surface of the earth, which is often turbulent and capped by a temperature inversion. ABL is characterized by strong vertical variations of temperature and wind velocity associated with significant exchange of momentum, moisture, heat and mass. The main physical feature of the ABL is turbulence that arises due to the instabilities induced by horizontal and vertical shears of wind close to the surface. The influence of surface friction and surface heating is quickly transmitted to the ABL by turbulent transfer.

The depth of ABL may vary from a few tens of meters to a few kilometers, depending on the surface heating, surface winds, terrain and roughness characteristics, vertical winds, horizontal advection of heat and moisture and other factors. When the boundary layer is neutrally stable its depth may be defmed as the height where the frictional effects are marginal. Under convectively unstable conditions, the top of ABL is capped by an inversion and under highly stable conditions the height of the inversion at surface gives the boundary layer depth. The depth of ABL may even up to 3 km in thermal convective conditions such as summer afternoons over India. The depth of ABL may be about 1 km in tropics under normal weather conditions. The height of ABL in the tropics differs with those in the middle and high latitudes due to smaller magnitude of Coriolis force and dominance of moist processes.

The depth of ABL differs over land and ocean, as its variation in space and time is slow over oceans. Over oceans, the surface forcing varies slowly due to small diurnal variations of sea surface temperature. Most of the changes in ABL over oceans are due to synoptic and mesoscale disturbances as upward motions in low pressure regions contribute to higher ABL depth. Generally this poses difficulty to determine the ABL height in which case cloud base is assumed to be the ABL height.

Although the ABL constitutes a small fraction of the atmosphere, the atmospheric conditions within the ABL assumes importance as it is the layer in which humans live most human activities takes place. Solar radiation, which is the primary source of energy for the earth- atmosphere system, is mostly absorbed at the ground and transmitted back to the atmosphere through the ABL. Turbulent transfer of latent heat and sensible heat is important for radiation balance and heat energy budget at earth surface. The turbulent transfer of momentum from the atmosphere to the earth's surface (sink for the atmospheric momentum) occurs through the ABL. Energy exchange processes in the boundary layer have profound influence on local weather. Boundary layer friction causes low level convergence and divergence of flow in low and high pressure regions. The frictional convergence in moist boundary layer air causes low level moisture convergence in low pressure systems. Even though the ABL comprises of a only small fraction of total kinetic energy of the atmosphere, about 50% of the atmospheric kinetic energy is dissipated in the boundary layer. Thus it is evident that the boundary layer processes affect the evolution of the weather.

The boundary layer has a well-defined structure. A very small thickness layer of few millimetres exists immediately adjacent to the earth surface, which is called “molecular sublayer” where viscosity and molecular diffusion processes control the vertical transports. Immediately above this layer is the “surface layer”, in which the variation of turbulent fluxes is minimal and generally has vertical extent of 10% of the total ABL. Above this layer is the “mixed layer” where the vertical gradients of mean properties are very small as a result of turbulent mixing. The turbulence in the mixed layer is usually convective driven hence it is also known as convective boundary layer. At the top of the mixed layer an inversion layer may exist which is the interface between the turbulent boundary layer and free atmosphere. At night time, boundary layer will be mostly stable as temperature lapse rate decreases. The height of the top of this stable boundary layer is usually taken as the height of the surface inversion.

The evolution of ABL is continuous responding to the heating and cooling of the earth's surface. From sunrise, the mixed layer depth increases, attains maximum a few hours after the noon time and has a capping inversion at the top. During daytime, the mixed layer is characterised by intense mixing due to rising of warm vertical air. After sunset, the capping inversion weakens and is replaced by a shallow inversion. During night, as the ground cools the air above it cools and mixes upward by turbulence generated by wind shear, and the surface inversion layer grows steadily to form a stable boundary layer. This sequence of boundary layer evolution is typical for land surface in the middle latitudes. In the tropics the base of the trade wind inversion serves as the top of boundary layer. The sub-cloud layer act as the inversion if clouds are present and the base of this sub-cloud layer is taken as the top of the boundary layer.

The ABL plays an important role in governing the atmospheric motions, as the energy exchange between earth's surface and free atmosphere takes place within this layer. The motions within the ABL are turbulent in nature, due to the aggregate effects of sensible and latent heat exchange at the surface and the generation of wind shear due to irregular terrain and surface characteristics. This is in contrast with the vertically stable state in the free atmosphere above the ABL.

The ABL is delineated by three sublayers with distinct vertical flux profiles and two different regimes identified by upward or downward fluxes of virtual potential temperature. The heat flux in dry conditions indicates the buoyancy flux that determines whether the kinetic energy of heated (cooled) parcels of air is changed by ascent (or descent) in association with turbulent eddies. The ABL is referred to as unstable or convectively active when the buoyancy flux in the surface layer is positive (upward ascent of relatively warm, buoyant parcels) and stable when it is negative.

The surface layer is the atmospheric layer nearest to the earth's surface, with a vertical thickness of a few tens of meters, where the vertical fluxes are nearly independent of height and satisfy the constraints of the Monin–Obukhov similarity. There is observational support for this from instrumented sites with meteorological towers. The vertical wind and temperature profile relationships that have emerged from analyses of these observations have provided the so-called flux profile relationships that have become the basis for parameterizing the surface fluxes.

The ABL is generally capped by a temperature inversion occurring depending on the strength of the turbulence in the ABL and its interactions with the free atmosphere (for e.g., the presence of subsidence in the free atmosphere above the ABL). In the region between the surface layer and the top of the ABL, turbulent eddies are deeper and produce mixing leading to well-mixed vertical profiles signified by weak vertical gradients of temperature, humidity and wind.

Upward buoyancy fluxes in the surface layer lead to a statically unstable (super-adiabatic) surface layer, which generates turbulence due to buoyancy production from strong vertical mixing and growth of the ABL. In a typical cloud-free daytime boundary layer, upward buoyancy fluxes result from surface heating due to absorption of solar radiation, and the vertical heat and buoyancy fluxes are upward between the surface and the base of the inversion, even though near neutral or slightly statically stable conditions prevail between the top of the surface layer and the top of ABL. In contrast, the surface layer becomes stably stratified at night when radiative cooling occurs at the surface and turbulent fluxes are downward and typically weaker throughout the ABL.

As noted in the preceding section, the effects of turbulent transfer in the ABL are represented by vertical and horizontal gradients of stress components and fluxes. The gradients of vertical flux and stress components are the most important for parameterization of ABL.

Equations governing these quadratic (second-moment) quantities can be derived but they involve cubic and higher order products that are not determined in closed form, but in general are not negligible in magnitude and must therefore be taken into account. Equations for these cubic quantities can also be derived but include higher order products that are also undetermined and so on. In principle, this hierarchy of higher moment equations is unclosed but in practice must be truncated by invoking closure assumptions. The problem of higher moment closure is central to modeling of turbulent flows. Traditional approaches that have been used in GCMs have evolved from the extensive body of knowledge and theory that has developed in this field over the past century.8 These typically involve representing turbulent transfer as being diffusive in nature, directed down local gradients of resolved variables with associated eddy diffusivities that are many orders of magnitude larger than the molecular viscosity and heat conductivity for air and usually depend on local dimensionless flow quantities such as the gradient Richardson number.

Traditional second-moment closure formulations for turbulent transfer that represent fluxes of heat, momentum, and moisture as diffusive transfers down local gradients of wind components, temperature, and humidity are not able to account for the typical vertical flux profiles in convectively active boundary layers without establishing states that are at least slightly statically unstable. Such states are inconsistent with both observations and simulations with high-resolution models that resolve at least the most energetic eddies, usually referred to ats large eddy simulation (LES) models. This limitation of local second-moment closure parameterizations has long been recognized. In these circumstances, the turbu-lence is convective in nature and the recognition that it is not strongly correlated with local vertical gradi-ents of temperature and moisture above the surface layer has stimulated a number of proposals for param-eterizing the nonlocal, counter-gradient heat transfer that is a salient feature of such convectively active boundary layers. There is not yet a well-established theoretical framework for these nonlocal parameteri-zations. There have been two basic approaches toward representing such effects. One of them attempts to account for counter-gradient transfer by developing equations governing second-moment quantities that include nonlocal effects through parameterizations of third moments in terms of second moments and mean quantities. This approach has been partially success-ful, particularly for cloud-free boundary layers, and is used in some GCMs.

A different approach, more recent and still not widely applied in GCMs, makes use of ideas that have traditionally been applied to cumulus parame-terization (discussed below) but adapted to modeling nonlocal transfer in the boundary layer. This approach postulates that the nonlocal transfer is convective in nature and associated with eddies that are deep enough to span the entire depth of the boundary layer or in some cases to overshoot the inversion layer and develop into cumulus clouds. The parameterization approach models these eddies as entraining/detraining plumes, wherein entrainment (i.e., horizontal flow into plumes from their environments) and detrainment (horizontal outflow from plumes) must also be param-eterized. More recently, there have been attempts to combine both approaches to achieve a more general approach. However, a unified theoretical framework for these approaches remains elusive.

Turbulent transfer in statically stable regimes is typically more local in nature and therefore local downgradient transfer formulations are more suitable for parameterizing turbulent transfer in the ABL in these circumstances. However, until quite recently, the theoretical formulations that underpin such parameterizations implied extinction of turbulence entirely at finite values of the local gradient Richardson number. A well-known limitation of using such parameterizations in GCMs is that they result in unrealistically weak downward fluxes in stable conditions and associated decoupling of the boundary layer and free atmosphere from the underlying surface. This may call into question the applicability of theoretical results including Monin–Obukhov similarity in the presence of strong stable stratification. There are also some observations that suggest that small-scale turbulent motions that produce mixing in the ABL may occur with relatively strong stable stratification. Recent developments in higher moment closure theory, combined with results of turbulence simulations using very high-resolution large-eddy simulation (LES) numerical models, have resolved this uncertainty and apparent conflict with Monin–Obukhov similarity to some extent. They have removed the restriction on the occurrence of turbulence associated with shear production in stably stratified ABLs that has hitherto been a characteristic limitation in GCM applications. Of course, it should also be understood that the horizontal resolution of most currently operational GCMs is so coarse that they are unable to adequately represent such effects as those associated with horizontal heterogeneity of the underlying surface, the intermittent and episodic nature of turbulence in the boundary layer, and coupling of turbulence with large-scale but still unresolved phenomena such as internal gravity waves.

Second-moment closure schemes are usually closed through introduction of turbulent length scales that characterize the sizes of energetic eddies. Apart from the law of the wall constraint that requires turbulent length scales to vary with the distance from surface in the ABL surface layer, there is abiding uncertainty concerning determination of turbulent length scales in the rest of the boundary layer and the free atmosphere. While some theoretical approaches to modeling the length scales have been proposed in the turbulence modeling community, these have not yet received wide application in GCMs. On the other hand, over the last two decades, various proposals to introduce nonlocal effects into modeling of turbulent length scales for ABL applications have been explored in the atmospheric ABL modeling community and some of these have been used in model applications. However, for the most part approaches to defining the turbulent length scales have remained ad hoc in nature.

Stratus and strato-cumulus clouds are ubiquitous in the ABL and interact with turbulence in number of ways that are important not only for pro-cesses within the ABL but also for interactions with the free atmosphere through radiative processes, coupling with moist mixing near the top of the ABL in association with both the effects of radiative cooling and evapora-tion of condensed water during turbulent entrainment events at the tops of stratiform cloud layers. When there is widespread cloudiness in the ABL, the tem-perature and humidity profiles that are characteristic of dry boundary layers are replaced by more compli-cated vertical structures that reflect the effects of latent heat release in the ABL. However, in the absence of significant amounts of precipitation, the total water (sum of water vapor and condensed water in clouds) should be conserved in principle and this has been found to be approximately so in practice for many ABL cloud systems so that the vertical profiles of total water are qualitatively similar to those for water vapor in cloud-free conditions. The latent heat release associated with condensation of water vapor typically couples with the turbulence in these circumstances to enhance the buoyancy production in the cloudy region of the ABL. However, approximate conservation of total water also implies approximate conservation of other thermodynamic variables that are combinations of temperature and water variables, such as the equivalent potential temperature. The vertical structures of these conserved variables in cloudy boundary layers are often similar to their counterparts in cloud-free ABLs.

Unrealistic modeling of ABL clouds has traditionally been a major shortcoming of GCMs. In recent years, substantial progress in the understanding and modeling of clouds in the ABL has resulted from organized observation and modeling programs such as those of the GEWEX (Global Energy and Water cycle EXchanges) Cloud System Study that have been directed toward improving ABL cloud parameterizations in GCMs. These efforts have relied extensively on comparing observations with results from high-resolution LES models, which resolve the main features of clouds, and with results from single-column models (SCMs) which are essentially stand-alone models derived from GCMs that include the main physical process parameterizations of the GCMs. These SCMs do not explicitly include processes such as horizontal and vertical advection. Therefore they are limited in their capability to produce closed feedback loops, which represent the complete response that is realized in GCMs when parameterized processes are included. However, they require much smaller investments of computing resources and it has been found that, when used in conjunction with LES models in the context of carefully designed observational case studies, they are able to reproduce key features of the response to parameterizations that is produced in their corresponding GCMs and therefore are valuable tools for improving the parameterizations that are used in GCMs.

5.4 Convection

Satellite pictures reveal that, at any given time, 60-70% of the globe is covered by clouds. Clouds are known to play a very important role in the earth-atmosphere system, to affect the radiative fluxes at the earth's surface as well as within the atmosphere, to contribute to heating due to release of latent heat (as water vapour condenses to form clouds), to contribute to vertical transport of moisture, momentum and heat and to contribute to the surface hydrology (as precipitation reaches ground). It is imperative that the impacts of the clouds on the evolution of the atmosphere are to be incorporated in to the weather prediction modeling system. Since individual clouds generally have spatial scales of few hundred meters, and the condensation and evaporation processes associated with cloud droplets have spatial scale of few micrometers and since it is not possible to explicitly resolve these processes with the model grids of the current models (~few kilometers in the horizontal and vertical), it is only possible to relate the structure of the clouds and the resolvable scale model variables through a closure assumption and statistical parameters. Different types of clouds exist in the atmosphere, such as stratocumulus clouds at the top of the convective boundary layer, upper tropospheric cirrus clouds, deep cumulus and cumulonimbus clouds etc., which are formed, maintained and dissipated by different physical processes such as convection, cloud microphysics, large-scale vertical motion etc. Due to their discriminating properties, cumulus and cumulonimbus clouds are being treated separately leading to the segregation of the treatment of vertical transport and condensation effects and the radiative effects leading to separate the parameterizations of cumulus convection and cloud microphysics. Thus the objective of the cloud parameterization is to properly portray the generation and dissipation of clouds and the precipitation formed inside clouds; and to furnish information of the cloud fraction area and cloud condensate for radiation parameterization calculations.

Observations indicate that the aggregate of all individual clouds is often denotes a small fraction of the grid area, and the parameterizations use this as a primary assumption. The clouds are assumed to form when the specific humidity value exceeds the saturation value, and the assumption of fractional cloud cover implies that some parts of the grid area become saturated earlier than the rest of grid box. The cloud fraction is decided, taken as zero below a threshold relative humidity (RH) value below and as one if the box is saturated. This cloud fraction concept also implies a distribution of the humidity from the saturation point within the model grid box around.

Next important part is the description of the condensation process. This comprises of two processes, one is the nucleation of cloud particles on condensation nuclei (small aerosol particles) when RH exceeds saturation or nucleation of large droplets under supersaturated condition and the other is growth of cloud particles by diffusion of water vapour. Generally, the presence of sufficient condensation nuclei is assumed to avoid complication of supersaturation conditions contributing to nucleation of ice particles.

At the beginning of the development of the atmospheric models, a simple scheme was used called "moist convective adjustment". In this scheme, supersaturated values (RH>100%) at any grid point are adjusted to saturation considering the increase of temperature due to latent heat release. Similarly, temperature lapse rates in saturated conditions in a grid column are adjusted to the moist adiabat. All the condensate is assumed to be the fall out precipitation. Early models assumed condensation process without radiative interaction and prescribed cloud albedo values were used for radiation calculations. The next advancement was to calculate cloud fraction based on RH (threshold taken as 80% usually) as described before, and in this scheme also the condensate effects were prescribed for radiation calculations. This approach is the "diagnostic" method as the cloud fraction and the condensate are diagnosed based on the grid point values. A further development was to include information concerning the vertical motion and strength of inversion at the top of PBL to derive the cloud fraction, which had been adopted till middle of 1990s as it provided reasonable estimates of the cloud distributions.

The next step was to add a prognostic equation for cloud condensate, which helped its direct inclusion into the radiation calculations, and to assume auto-conversion of a part of the cloud condensate as precipitation.

Along with, another approach based on using probability density functions to define the distributions of the temperature and humidity with respect to grid-scale mean values was developed and used to define the cloud fraction and condensate within a grid element.

In 1990s, fully prognostic models have been developed, in which equations for the time change of cloud condensate and fraction are dependent on the processes of advection, generation and dissipation, and this method had been widely used. The drawback of diagnostic cloud estimation in these methods lead to the development of process oriented approaches, in which the grid-resolved large-scale vertical motion as well as the parameterized convection directly affects to change both the grid-scale variable and the cloud variables. This approach is complex as the effects of each of the physical processes on the clouds are to be explicitly stated.

At the current time, mass flux based approach for convection parameterization is the most widely accepted and adopted. In this, two important issues that need to be resolved are of the closure and the entrainment/ detrainment. The basic assumption, in this method, is plume type convection such as cumulus towers. The cloud base mass flux, (M,) is the key variable that needs to be determined. Assuming this as known, the budgets for mass, heat and moisture at different levels could be determined using standard thermodynamics. Further, assuming the rates of entrainment and detrainment, the vertical profile of the mass flux could be determined. A closure of the problem is achieved through an assumption such as "quasi-equilibrium" between large scale and convection. Several investigations were made to assess the impact of entrainment and detrainment assumptions on the production of shallow and deep convection, which have finally concluded that closure is more important. In the closure hypothesis, both the processes within the boundary layer and the troposphere seem to be important but some studies have brought out large-scale forcing (upward motion) impacts the formation of convection more than the boundary layer forcing as the boundary layer processes are noisy and does not support slowly evolving mesoscale convective systems. The concept of convective quasi–equilibrium was postulated by Arakawa and Schubert in 1974,

Arakawa and Schubert's convective quasi–equilibrium, postulated as a balanced state in the cloud work function budget, between the large scale forcing supporting atmospheric instability and the convection supporting stabilisation of the large scale environment. This concept could be verified with observations and also had the possibilities of extending the hypothesis.

It is known that convective parameterization (CP) schemes are formulated to estimate the vertical transport of latent heat, and redistributing the temperature and moisture in a grid column leading towards stabilisation of the atmosphere. Any of the CP schemes, therefore should include the triggering function for convection, processes to modify the grid column and the mutual interaction of the grid-scale and convection processes.

The characteristics of some of the schemes are briefly presented here. Early schemes were mostly empirical, with emphasis on distributing the heating rates and precipitation. Later schemes are based on well-defined physical process representation such as mass-flux or convective adjustment hypothesis.

5.5 Kuo Scheme (Empirical)

This is a simple scheme that adjusts the vertical temperature and humidity profiles to moist adiabat and produces precipitation. In this method, convection is triggered when the vertical column-integrated moisture convergence exceeds a threshold value.

As the temperature profile adjusts to the saturated adiabat through the cloud, some of the moisture is assumed to moistening the environment and the rest is assumed to condense as precipitation. So the precipitation produced may be different, even for the same sounding, due to variations in the assumption on the partitioning towards moistening and precipitation.

The duration and intensity of convective precipitation depends on the low-level (boundary layer) moisture convergence. The vertical profiles of temperature and moisture only tend towards the moist adiabat as the scheme assumes that a fraction of grid is only covered by convection. The advantages are only that it is simple to understand its process application and runs faster with lesser computational resources. The limitations are that it cannot account for inhibition of convection and that it can trigger repeated convection as heating from the parameterization supports moisture convergence.

5.6 Betts-Miller-Janjic Scheme

This is a convective adjustment scheme, in which the model vertical profile is subjected to adjust to a pre-defined reference profile. The reference profile is determined from climatology of the profiles corresponding to post-convective stage with representative values taken at the cloud base, cloud top and the freezing level. Different reference profiles can be used as representative of seasons and tropical and extra-tropical regions. There are three conditions to trigger the convection, which are a minimum CAPE, convective cloud depth higher than a threshold value and active moist sounding.

In this scheme, rain is produced with reduction in precipitable water as the profile gets adjusted to the reference sounding, keeping the consistency of warming with the latent heat release due to condensation. This means that the profile of sounding vacillates around the reference profile till the latent heat released due to precipitation process is consistent with sensible heating of the sounding. In this scheme, horizontal and vertical advection in the large scale flow controls the moisture depth and instability and modifies the vertical column by reducing instability. This scheme adjusts the vertical column to the reference profile, does not modify the sub-cloud layer as there is no cooling from downdrafts, however cooling of the sub-cloud layer may be possible due to evaporation cooling of falling precipitation and a reduction in the incoming solar radiation. This scheme works well in moist environment, implicitly includes the effects of downdrafts, latent heat of fusion on clouds and is known to be the best scheme that prevents the impacts cloud microphysics scheme to force convection.

This scheme has the limitations of non-suitability for dry arid regions, reference profiles are climatology based and not flexible, induces precipitation very quickly, does not impact atmosphere below cloud base as the effects below the cloud base are indirect through reduction in insolation or cooling due to evaporation of falling precipitation.

5.7 Arakawa- Schubert Scheme

This is the most realistic physically based scheme, though complex. The physical processes of moisture detrainment from convective clouds, warming due to environmental subsidence, and quasi-equilibrium between convective stabilization and large-scale destabilization are explicitly included.

The triggering of convection is complex as it is dependent on the boundary layer energy, the large scale destabilisation rate and the cloud entrainment at different levels.

The basic assumption is that the clouds area constitutes a small fraction of the grid area, implying that cloud temperature and moisture does not affect the grid variables. The rest of them are that: clouds with different heights are assumed to form, with their bases at the top of the boundary layer, and with different tops dependent on the entrainment rate (no entrainment producing the deepest cloud); compensating subsidence occurs within the grid contributing to warming and drying; downdrafts detrain below the cloud base contributing to changes in the boundary layer; and detrainment from the cloud is only at the cloud top contributing to cooling and moistening at that level.

The properties of cloud updraft at different heights are obtained using a one-dimensional cloud model, which are further used to calculate the temperature, moisture, environmental subsidence, amount of cloud air detrainment into the environment, and the amounts of precipitation evaporation contributing to a downdraft and precipitation falling to the ground.

As the environmental changes take place due to the combined effect of detrainment from clouds at their cloud top levels, environmental subsidence, and boundary-layer stabilization due to convective downdrafts, and that some of these contributions offset each other, the ultimate changes are small. The one dimension cloud model does not reflect the three-dimensional cloud structure and although confined to parcel buoyancy lifting process only, it includes many of the important processes such as entrainment, detrainment, evaporation, buoyancy parcel mixing and precipitation generation.

This scheme uses a novel idea of "quasi-equilibrium" between grid-scale forcing and the resulting convection. Convection thus formed compensates for CAPE and controls its enhancement. This scheme impacts to cooling at lower levels and warming at upper levels leading to stabilisation. The affected changes in the atmosphere are small and any small destabilisation due to advection or radiation may trigger more convection. However the model vertical sounding may be very much different due to other dynamical and physical processes. The AS scheme allows the simultaneous existence of different cloud types, deals with the processes of entrainment, detrainment, compensating subsidence associated with the clouds and is found valid for many weather situations.

The limitations are insufficient stabilisation from clouds leading to more grid-scale convection; assume all clouds to have their base at the top of the boundary layer and so does not consider elevated convection; assumption of convection area to be small as compared to grid area which is questionable in the current context of high resolution models; and the assumption of entrainment in to clouds from sides in contrast to observation of the entrainment at cloud top level in deep clouds which may lead to errors in the estimation of precipitation and heating profile. Above all, this scheme takes more computational time than many of other schemes.

In NWP models, parameterizations of convection and cloud microphysics are separately dealt with. Microphysics part concerns with the precipitation and cloud processes as excess moisture is removed considering the grid-scale temperature, moisture and wind variables, which are adopted on grid-scale. Convection parameterization concerns with the process of convection generation that affects to reduce the atmospheric instability through redistribution of vertical temperature and moisture profiles in a grid column, which tends to reduce unrealistic grid-scale precipitation.

As the model resolution increases, more detailed information on the microphysics inside clouds is required so as to include more of the hydrometeor information, and the microphysics information to be shared with convection parameterization schemes.

The interaction between the CP and microphysics are detailed as follows: In an initial unstable environment, the CP scheme produces updrafts and downdrafts and precipitation affecting reducing the atmospheric instability and modifies the vertical profiles of temperature and moisture. The cloud scale updrafts lead to heating and moistening in the middle and upper levels which further leads to increased grid-scale vertical motion. Thus meso-scale circulations are generated as driven by heating due to latent heat release at higher levels and cooling at lower levels due to evaporation. These changes in the heat and moisture distribution will lead to generation of more grid-scale precipitation and hydrometeors. In contrast, in an atmosphere which is initially stable, clouds are first formed from water vapour followed by the formation of cloud liquid and ice and their fall out.

As model resolution increases, the processes of convection and cloud microphysics can be treated together. At some high resolution, CP may not be required and their effects could be included explicitly. Since the vertical and horizontal motions are of the same scale in convective systems, vertical accelerations become equally important or sometimes dominant which prohibits the use of hydrostatic assumption and requires the application of nonhydrostatic models. It should be noted here that even with 1-2 km grid interval, the resolution of cloud structure will still be crude and unrealistic.

5.8 Cloud Microphysics

Parameterization of grid-scale cloud microphysics concerns with the cloud and precipitation processes, which describes the removal of the excess moisture directly due to large scale forcing such as convergence or orographic lifting. It is important to note that, although grid-scale motions force to influence the cloud and precipitation, subgrid-scale cloud processes also have to be considered to fully describe the cloud microphysical influence.

The development of clouds and precipitation in the microphysics scheme contributes to release of latent heat which modifies the wind, temperature, and moisture fields. Atmosphere below the cloud base layers will be subjected to cooling due to evaporation of falling precipitation. These processes will strengthen the circulation responsible for clouds and precipitation as feedback.

Simple cloud schemes determine precipitation from cloud water or ice only, whereas schemes with prediction of clouds follow a sequence of physics-based events before yielding precipitation. Schemes that use complex models will predict precipitation considering physical processes for different hydrometeors and modeling of the internal cloud processes.

5.9 Simple Cloud Schemes

These schemes first determine the sub grid-scale moisture variability using a threshold relative humidity (generally taken <100%) and then the amount of cloud water (ice). Clouds are assumed to form and the cloud fraction is estimated depending on the RH value above its chosen threshold value.

The simple cloud scheme has the advantages: possibilities of comparing the model cloud fields with satellite imagery; possibilities of assimilating cloud data; direct relationship between cloud and radiation processes; distinction of cloud water and ice help radiation calculations; predicted relative humidity fields are more realistic due to allowance for simultaneous existence of water and ice; and better linkage of microphysics with convection parameterization. The disadvantages are that precipitation is a derived product (not predicted directly); hydrometeors are not explicitly predicted; computed precipitation rates are grid-box averages and so could be over or under estimates; and precipitation rates may vary within the grid area which manifest as its increase with larger grid area.

5.10 Complex Cloud Schemes

These schemes are complex and suitable for high grid resolutions that are capable of resolving small-scale circulations that influence microphysical processes. The process of modifying the environmental RH will be affected through four steps as follows. First step is to use a threshold RH value (<100%) to determine the sub grid-scale moisture variability, assuming that supersaturation is not required for liquid and ice formation and to include cloud processes for mixed phase and graupel. Second step is to induce environmental warming due to condensation process leading to liquid, ice or any hydrometeor and thereby lessening the environmental specific humidity. Third step is to affect process for cooling and moistening at the freezing level (due to ice melting) and below the cloud base (due to evaporation from falling precipitation), and to account for phase changes and hydrometeor interaction. Finally, precipitation is estimated considering moistening and cooling of sub-saturated regions as it falls to the ground and also retaining some amount of water and ice in the clouds all to produce a realistic RH in the grid column.

These models have the advantages of predicting the amounts of each type of hydrometeor that reaches the ground directly; prediction of graupel; predicting the cooling due to evaporation or melting of precipitation; better representation of environmental RH fields due to the property of retaining a part of water or ice in clouds; capability to assimilate satellite-based hydrometeor data, radar data; prediction of aircraft icing based on available super-cooled water in clouds; and better cloud-radiation interaction due to hydrometeor prediction.

The disadvantages are the requirement of increased computer time and memory; requirement of high resolution that could resolve small-scale features; difficulties to decide which of the hydrometeors are important for certain situations (such as aircraft icing); spin-up problem to achieve equilibrium between the initial values of hydrometeors and model temperature, moisture and wind variables (although could be reduced by considering guess fields of hydrometeors).

The cloud microphysics scheme tends to remove atmospheric instability by producing grid-scale convection, which in turn modifies the vertical profiles of temperature and humidity.

In the simulation of clouds without applying convective parameterization (CP), the model vertical motion (due to convergence) will moisten the atmospheric layers upward. As the updraft contributes to saturation of the entire grid-box, microphysics will produce precipitation. As the grid-scale vertical velocities are very small than the convective updrafts, the condensation process is slow and effects moistening of the lower levels.

CP schemes parameterize the intense vertical transports done by sub-grid scale updrafts and downdrafts, carrying the diabatic heated air in the lower troposphere into the upper troposphere and evaporation cooled air in the mid-troposphere into the boundary layer, and the compensating subsidence causes environmental warming.

6. Weather Prediction – Lorenz Chaos Theory – Nonlinear Dynamics, Ensemble Prediction

Any scientific method would follow the observations of phenomena, analysis of observations to classify them and make deductions and further use the deductions for making predictions. Finally the predictions are compared with observational characteristics of the phenomena for validation. Mathematics is essential to the physical processes, as observations are expressed as numbers, deductions are expressed as mathematical equations and these equations are used to make predictions. Some preliminary definitions are that a "system" is something that is to be predicted; the "phase space" is considered as a representation of the possible states of the system; the phase space uses numbers to define the system state; and the number of variables needed to represent the system is called its "dimensionality".

The world is not static but dynamic meaning that it consists of systems which change with time. As a system changes, the numbers representing the state of the system in the phase space also change. A dynamical system is one in which the phase space and the representative numbers change as per the governing rules. Generally the changing pattern of the phase space is called an orbit. So, a dynamical system means that its future state could be completely predicted from its current state.

Time is continuous and if the system evolves continuously, the time evolving dynamical system is called a flow. The study of these systems uses calculus. Let $x(t)$ represent the state of variable x at time t, and the flow is described as $[dx(t)/dt]= F[x(t)]$ where $F[x(t)]$ is the function governing the evolution of the system, and this type of equations are called "Ordinary Differential Equations".

A linear system is a dynamical system whose evolution is a linear process. A linear process is one in which, if a change in any variable at some initial time produces a similar change in the same or some other variable at some later time. All systems that are not linear are called nonlinear systems, in which the change in a variable at an initial time can lead to a change (in the same or a different variable) at a later time that is not proportional to the change at the initial time. For a linear system, two solutions can be combined which will also be a solution for the system and defines the linear systems to be tractable. This is not true for nonlinear systems, which cannot be solved separately and so is to be considered as a complete entity. Nature is inherently nonlinear. Many of the nonlinear systems are approximately linear for small perturbations around the mean. It is known that nonlinear problems are not exactly solvable, and understanding the behaviour of the nonlinear systems was not possible before the advent of the computers. The nonlinear systems are present everywhere in nature and since their behaviour is different from the linear systems, understanding of the linear systems does not help the solution of nonlinear systems. Nonlinear systems have variety of characteristics including chaos, which reflects our notion of determinism.

Chaos means complete disorder and confusion, (i.e.) stochastic (random) behavior, whereas a deterministic system is governed by exactness without an element of chance. This raises the question as how a deterministic system can have chaos.

Lorenz (1963) considered a simple model in three dimensional space, with twelve variables (twelve dimensional space), to show the weather prediction as sensitive to initial conditions and postulated the "butterfly effect". This means that the motion will be in a bounded region of phase space, where a particle orbit eventually gets closer to its earlier position and again moves away. If there is no butterfly effect, the orbit will stay close to itself and the cycles repeat with a regular behavior. Due to the butterfly effect, the orbit never comes close to repeating itself leading to apparently random behavior, called "Chaos".

What are the limits to prediction? The impediments on the acquisition of observations with infinite precision put a limitation on the prediction of future state of a chaotic system. In a chaotic system, small errors grow exponentially with time, which means that the required precision on the initial conditions should also increase with the increase of prediction time. For e.g., if the tolerant errors are of 10% for 1-hour and 1% for 2-hours prediction, then errors should be of 0.001% for 4-hour prediction. Coming to weather prediction, with the availability of very good models to represent the weather evolution and along with the observational information, it is to be realised that it takes about 2-weeks time for a disturbance of 1-kilometer spatial scale to grow to global scale weather pattern, and so the effective limit of weather prediction is about a week. This leads to the inference that the weather system is chaotic. Lorenz commented that this situation can be understood as that the flappings of a butterfly could be responsible for the occurrence of a cyclone at some place on the globe within a few weeks of time.

The general philosophy of science is Laplacian determinism, meaning simple rules signify simple behaviour and complex rules are required to connote unpredictable behaviour. This negates the idea of simpler rules imply simple systems and lead for reconsideration of determinism. This implies that when all the equations governing a system are known, the science can lead to its future prediction exactly. Lorenz chaos theory suggests that simple rules can lead to complex behaviour and random behaviour of the system, and chaotic systems are sensitive to initial conditions. This sensitivity will put an effective limit on its predictability.

Lorenz, being a meteorologist interested in weather forecasting, was concerned in the predictability of the weather through solving the hydrodynamic equations. He performed a simple numerical experiment, in which the fluid motion is governed by thermal buoyancy (convection) and the model system is a set of simple equations whose solutions represent a deterministic nonperiodic flow. Lorenz conducted an NWP experiment inadvertently, by starting the prediction using the output from an earlier prediction experiment as a continuation of the earlier prediction. From the results, it was realised that the predictions from the original experiment and the subsequent one were different and that the differences arose due to rounding off of the output from the first experiment at 3 decimal places whereas the computations in the first experiment were continued at 6 decimal places (approximately 1% error). Thus, Lorenz found that small differences in the initial conditions (as attributable to the magnitude of the observation errors) may lead to divergent predictions. This led to the formation of the theory of chaos through a simple numerical prediction experiment using three variables in a non-linear system.

Lorenz defined the problem in a simple way, writing down the equations for thermal convection in a thin fluid. As thermal convection means rising of hot air and sinking of cold air, the problem is classic as it defines the development of convection in a thin layer of fluid depending on the amount of heating from below. As known, the fluid that is hotter than its surroundings will try to rise and will lose heat due to cooling. If the cooling is faster, then it will remain in its place, or if it loses heat slower than cooling it will rise up (convection). When the convection is confined to a thin layer, the patterns could be very different from nil flow to simple rolls to turbulent chaotic flow.

Lorenz assumed a simple model of thermal convection, assuming that the fluid velocity could follow a single roll and the temperature could have steady state solution and so have two time dependent modes. One of the variables is 2-dimensional velocity W (which is a function of time), positive value indicates positive roll (i.e.) clockwise spin and negative values indicate anti-clockwise spin and higher values of W meaning faster spin. The second variable is temperature, of which T_1 (t) and T_2 (t) describe the structures of the horizontal and vertical temperature fields. T_1 is positive (hot) to the left and cold to the right of the fluid box. T_2 is always positive and higher values mean that the top of the fluid is warmer. If both T_1 and T_2 are zero, the temperature is layer stratified and with hot fluid at the bottom and cold fluid at the top. This description means that the initial state is well defined and known and the future values of W, T_1 and T_2 are to be known.

Lorenz defined the equations for a dynamical system with the three time dependent variables as

$$\frac{dW}{dt} = \sigma(T_1 - W)$$

$$\frac{dT_1}{dt} = -WT_2 + rW - T_1$$

$$\frac{dT_2}{dt} = WT_1 - bT_2$$

The first term in the temperature equation represents the conduction of heat (no fluid motion) and r is the temperature difference across the thin layer. The mode structure will be sinusoidal in both the horizontal and vertical directions. The width of the system is one wavelength. The numerical coefficients are chosen to simplify the dynamical equations. The parameter σ, which depends on the properties of the fluid (i.e. the ratio of the viscous to thermal diffusivities, 1-4 for water; 0.7 for an ideal gas; and 10-1000 for oils) and the number b (which varies with wavelength) are taken as constants with values of σ=10 and b= 8/3, whereas the Rayleigh number (temperature difference) r is taken as a variable. The Rayleigh number 'r' is a measure of heating of the fluid (If r <1, the layer is being heated too mildly and so no convection and r >1 will have differing solutions as larger values of 'r' indicate stronger convection). The salient features of these equations are: (i) The time variation terms do not appear on the right hand side; (ii) They involve only first order time derivatives such that the evolution depends only on the instantaneous value of W, T_1 and T_2; (iii) They are nonlinear, because of the quadratic terms WT_2 and WT_1 in the second and third equations; (iv) They are dissipative, as $dW/dt = -\sigma W$ correspond to decaying motion, and (v) The solutions are bounded.

The solutions to the system of equations, as 'r' increases (i.e. as heating increases) are shown in Figure 6.1. The left hand side part of the figure shows the time series of all the 3 variables, and the right hand side shows the modal structure of the temperature field. For small values of r <1, the initial temperature perturbation cools faster and inhibits convection.

At r =1.1, the convective roll spins clockwise with a constant velocity (due to hotter left hand side, and the roll would be anticlockwise if the initial hot region is to the right side of the fluid). As the r number is increased, till r <24 the initiated convection will ultimately settle down to a steady clockwise or anticlockwise spin and as the Rayleigh number increases it

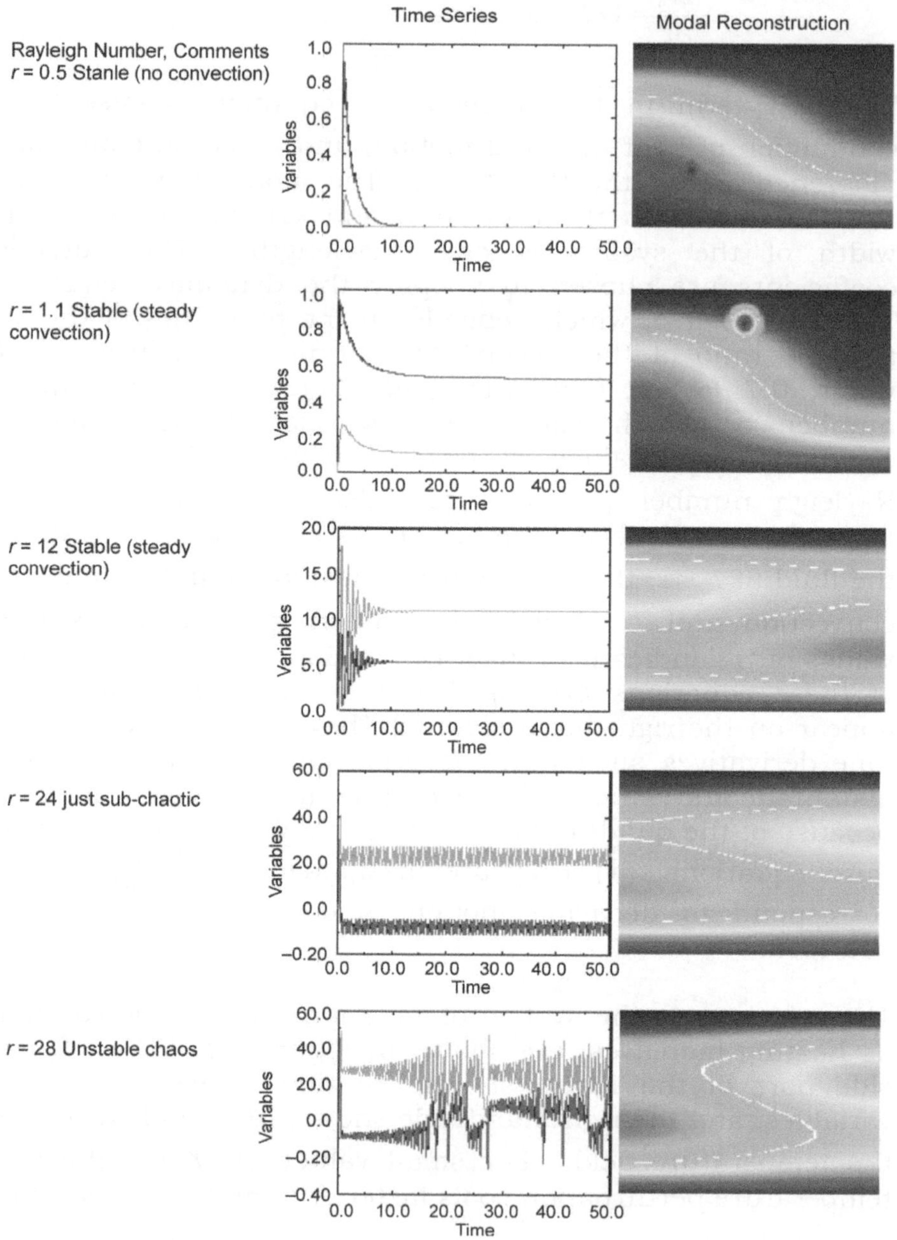

Figure 6.1 Behaviour of Lorenz equations as a function of Rayleigh number 'r', with constant values of b = 8/3, σ = 10. Initial conditions are W = 0; T_1 = 0; T_2 = 0, and with the left side of the box slightly warmer than right.

takes longer time to settle down to the steady state. At a critical value of the Rayleigh number r (identified as 24.74 in this case), the convective roll had become unstable with continuous oscillation of the roll in time with a flip of predictable direction but in an unpredictable way. This is "Chaos". The result from this experiment is stunning because the equations are deterministic (as the equations would describe the time behaviour of the three variables exactly) but the final solution could become unpredictable as the equations are integrated for a long time. Thus, Lorenz had made a great revelation that the earlier concept of future predictability with an exact mathematical system is not true.

This peculiar result is called deterministic chaos, as the basic problem is deterministic (as the short-term result is predictable and so not random) but at the same time chaotic because small errors in initial conditions lead to very different results due to nonlinear growing errors in long-term prediction. Lorenz interpreted this as "*butterfly effect*", as a flap of butterfly at one place makes the weather prediction beyond a week implausible. However, Chaos does not mean unpredictability. The behaviour of Lorenz system has shown that the solutions have some pattern, although doing flip and flop, to fit into a pattern of "strange attractor" (which means that the system cannot suddenly stop or start) that repeats quasi-periodically. So, the behaviour has some pattern or order but cannot be predictable over long time periods.

Lorenz, realising that the pattern could be followed for some definite period of time from the deduced results, had shown that the behaviour could be predicted at least one or few cycles ahead. Lorenz posed a question as "how far into the future one can predict from the information contained in a time series. He formulated a time series of the peaks of a variable, and tried to predict the next peak value from the earlier time series (i.e.) a list of peaks was prepared and for peat at time 'i+1' has been obtained as function of peak at the time 'i'. His results are shown in the Figure 6.2, which indicates skilful prediction of the next peak one cycle ahead. Repeating this exercise with varying lead cycles, leads the predictability into future. This is "order in chaos" and the basic inquest of nonlinear forecasting. This experiment shows that low dimensional chaos could be predictable longer than expected.

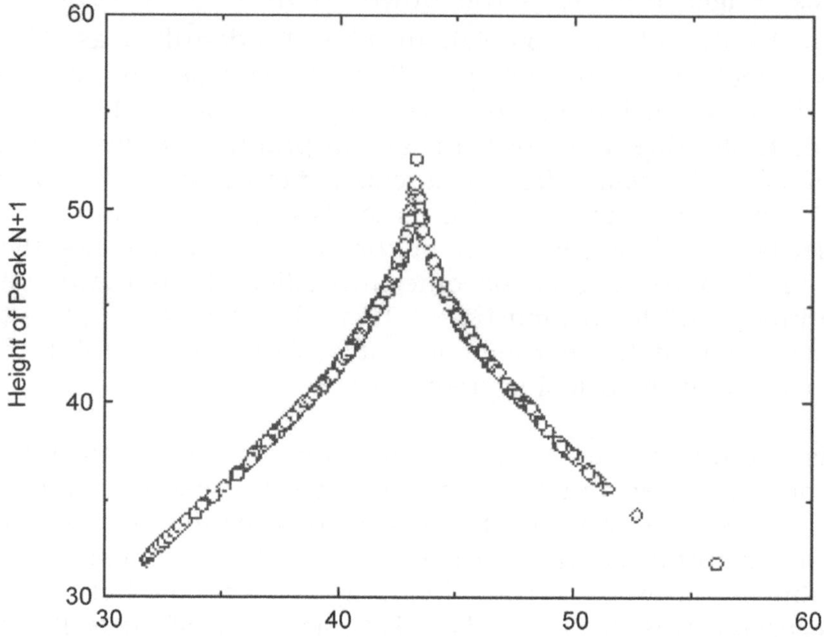

Figure 6.2 The Lorenz Map for predicting the height of the next peak given knowledge of the current height. This graph suggests that the height of the next peak can be predicted well.

The trajectories of the solutions of the nonlinear equations used by Lorenz, show patterns which seem to be deterministic but are subject to abrupt and random change which are referred to as "Lorenz attractor". Lorenz model indicates that the predictability of a chaotic system is flow dependent, meaning that some weather systems may be predictable and others may be highly unpredictable and also that the predictability itself varies and is predictable (Figure 6.3). The upper panel Figure 6.3 shows the trajectories from two different initial conditions, after time steps of 2, 4 and 6. The results of the control and perturbed forecasts are shown in red and blue respectively. It is interesting to note that the starting points of the two forecasts are negligibly separated (~0.1%, a representative value for observation errors); nearly same after 2nd time step; start to diverge after 4th time step and significantly divergent after 6th time step. At this time, the two patterns two different regimes. These figures brings out the divergence of forecasts within a short time from the start of the model integration, manifestation of two favoured regimes and a region between the two regimes from where the start of model integrations will have large uncertainty. The bottom panel show the pattern of trajectories from Lorenz experiment in 3-dimesions of T1, T2 and W and in the planes of T1-T2 and W-T2. This result is important in weather prediction as it shows the possibilities of assessing the probability of the predictions. In an earlier study, Lorenz demonstrated that errors at cumulus scale can lead to errors in synoptic scale within two days and the global scales within two weeks, and the result assumes importance in this context.

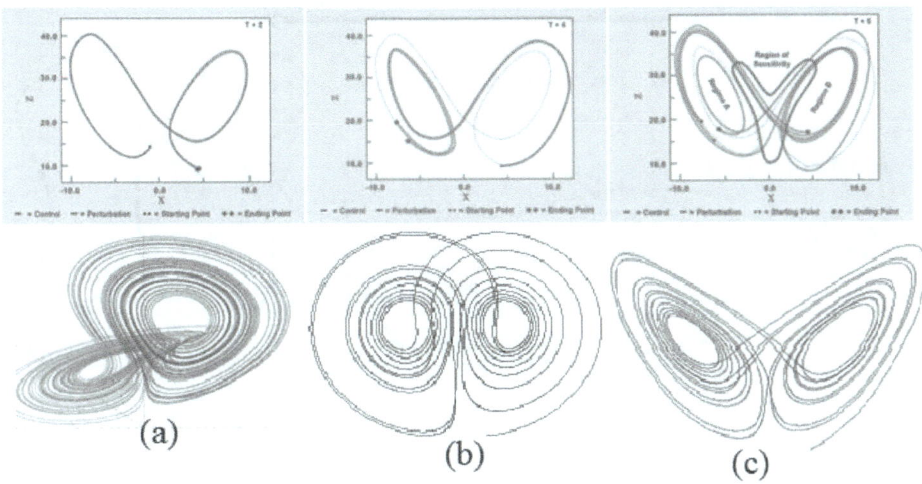

Figure 6.3 Upper panel denotes the particles' trajectories at different time steps of T+2, T+4 and T+6. Bottom panel shows the patterns of Lorenz attractor: (a) 3- dimensional pattern; (b) pattern in T1-T2 plane; (c) pattern in W-T2 plane.

Lorenz has shown that the uncertainties in the initial condition, however small they are, would lead to uncertainty in the forecast after a certain time period depending on the initial state. Due to this reason, numerical weather prediction studies have started using probabilistic methods to extend the deterministic limit of prediction beyond a week.

These probabilistic methods are based on using small random perturbations to the atmospheric variables, and these predictions (due to induced perturbations) would diverge (from each other) due to nonlinearity and chaotic process of the model atmosphere [Figure 6.4]. This procedure is now called "ensemble prediction" and each of the prediction from the perturbed conditions is called a 'member' of the ensemble. An example of the advantages of ensemble prediction is shown in figure 6.5. As seen, the deterministic prediction did not reveal the intensification of a weather system, whereas majority of the ensemble members have produced the intensification. By now, the probabilistic weather prediction has become common, providing a number of plausible weather predictions that could be used to determine the probabilities of occurrence. These probabilistic predictions would provide an opportunity to assess from the assessment of uncertainties.

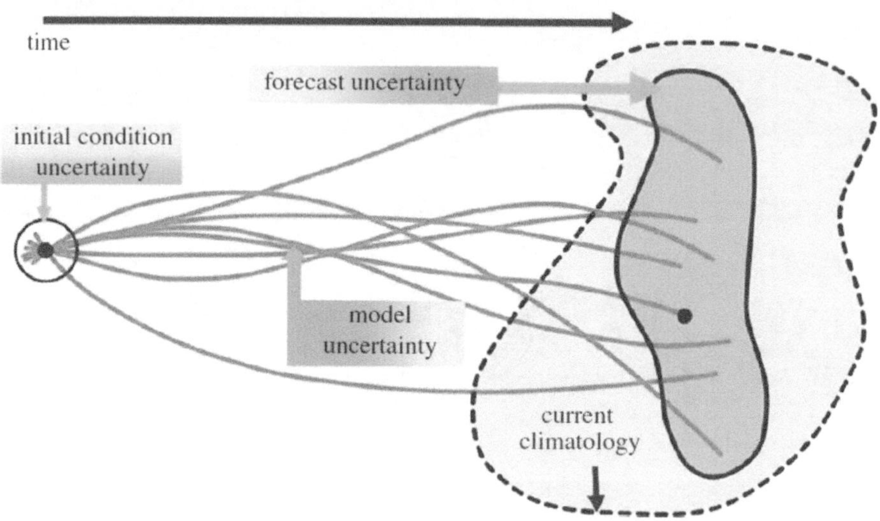

Figure 6.4 Probabilistic weather forecast using initial condition uncertainties. The blue lines show the trajectories of the individual forecasts that diverge from each other owing to uncertainties in the initial conditions and in the representation of sub-grid scale processes in the model. The dashed, lighter blue envelope represents the range of possible states that the real atmosphere could encompass and the solid, dark blue envelope represents the range of states sampled by the model predictions.

However it is important to note that the usefulness of the ensemble prediction is dependent on the representation of the uncertainties in prediction. It has been realised that the use of perturbations to the initial conditions may not always be enough dispersive as evinced by the increase of mean root square error growing faster than the spread (Figure 6.6). This lack of spread indicates the insufficiency of the perturbation based ensemble, which indicates that something else that is responsible for the uncertainties is not known or identified.

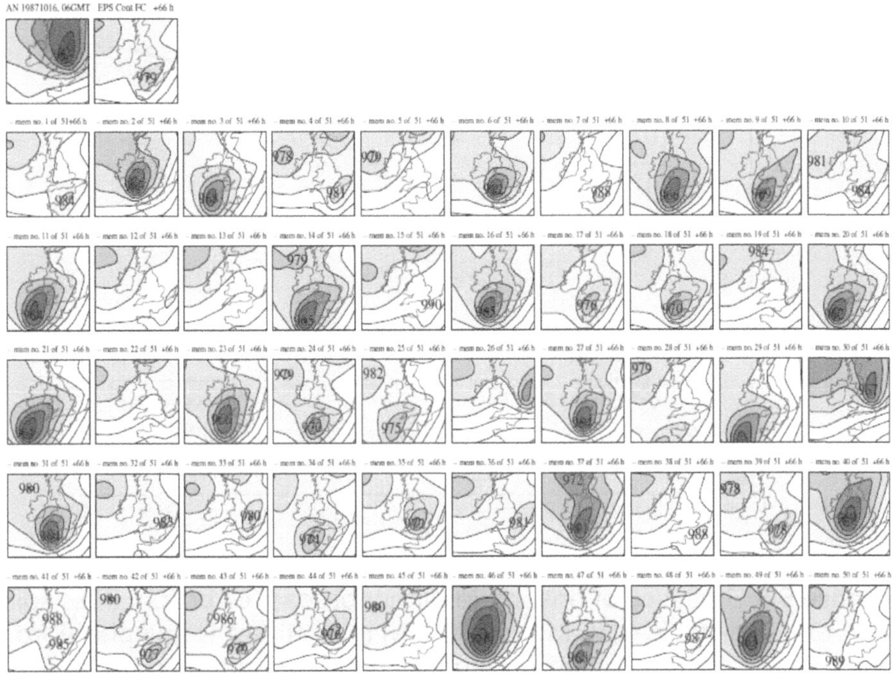

Figure 6.5 Example of 66 h probabilistic forecast for 15–16 October 1987. Top left shows the analysed deep depression with damaging winds on its southern flank. Top right shows the deterministic forecast, and the remaining 50 panels show other possible outcomes based on perturbations to the initial conditions. A substantial fraction of the ensemble indicates the development of a deep depression [Slingo and Palmer, 2011].

Several numerical studies in recent years have brought out that the uncertainties in the predictions could be related to the model formulation and the parameterization of physical processes. As it is known, empirical assumptions based on statistical properties are being used in the representation of subgrid-scale processes although observations suggest a broad range of formulations. Thus, it is understood that the model uncertainties and systematic biases restrict the sampling leading to under-dispersion of the ensembles. Some studies performed using perturbations in the closure parameters of the parameterizations indicated some improvements in the ensemble spread.

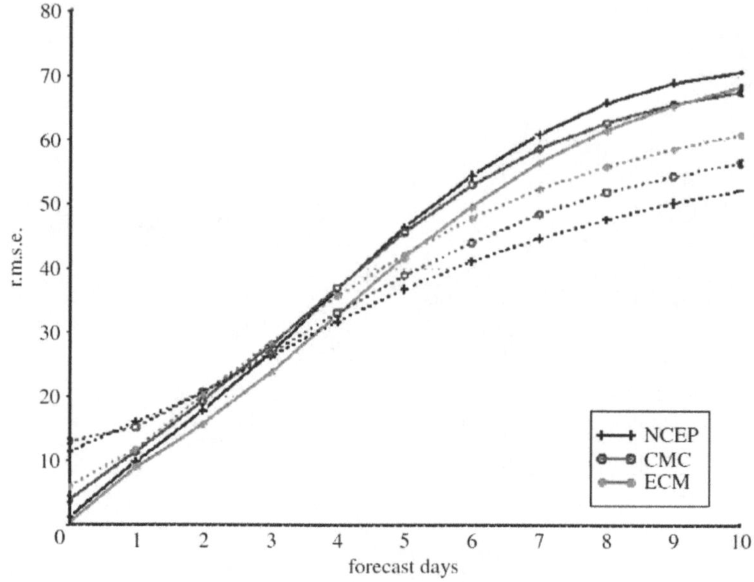

Figure 6.6 Statistics of ensemble mean forecast error (r.m.s.e.; solid line) and ensemble spread (dotted line) in Northern Hemisphere from three ensemble prediction systems. The systems based on random perturbations to the initial conditions were not sufficiently dispersive as the root mean square error grows faster than the spread, which indicates that the ensemble is under-dispersive and hence the ensemble forecast is over-confident.

The above narration relates to the Lorenz chaos theory in the context of weather prediction due to nonlinear dynamics inherently present in the atmospheric system. The limits to predictability in the deterministic procedure are understood as due to uncertainties in the initial conditions and the currently used ensemble approach that would enable probabilistic weather prediction and with increasing time scales.

To summarise, modern meteorology is concerned with the nonlinear dynamics of the atmosphere and ocean (ensemble prediction, etc.). Although a great deal is now known about low dimensional chaos (the erratic-like behaviour of dynamical systems described by a few variables), much less is understood about spatio-temporal systems with a large number of degrees of freedom, as those describing the atmosphere-ocean evolution. The nonlinear effects that drive the dynamics of finite perturbations are a key factor for the understanding of error growth and ensemble prediction in these systems. However, most of the methods currently used in meteorology rely on findings from the low dimensional world and do not take into account the interactions between space and time inherent of these systems.

Ensemble prediction

The weather prediction experiments of Lorenz, which lead to the theory of chaos brings out the limitations of single deterministic forecasts. The diversity of forecasts that could emerge due to errors in the initial conditions, approximations in the parameterization of the physical processes and the model formulation need to be well understood so that the inherent uncertainties that exist in numerical weather prediction are known to the forecaster. Further it is equally important that the public should be conveyed of the same in terms of probabilistic predictions. These probabilistic predictions appraise the understanding of the atmospheric behaviour and its modelling strategies.

In the NWP system, although it is desirable not to have errors, initial and model errors are unavoidable. After Lorenz chaos theory, which emphasized the growth of errors in NWP, ensemble prediction system evolved to enable forecasters to provide probabilistic forecasts along with a single deterministic forecast.

Firstly considering the inaccuracies in the initial conditions, the strategy of producing probabilistic predictions prompts the generation of a finite set of possible predictions that would emanate from different initial conditions at the same time point. This implies that the estimated initial conditions of the deterministic forecast are subject to a known range of error, which technically provides infinite sets of initial conditions. However, practicality restricts to the generation of a finite number of sets of initial conditions within the range of variability and conforming to the dynamics. These sets are referred to as ensemble members and the predictions as "ensemble prediction". In the deterministic single prediction method, use of data assimilation would minimise the errors in the initial conditions (as a part of the uncertainty is accounted for through imposing weighting of observations and background field (i.e.) first guess from model forecast) but cannot provide their impact on the future prediction. In contrast, ensemble predictions provide information of the impact of uncertainties in initial conditions by relating the fast growing errors to the perturbation in the initial conditions. An analysis of the development of forecasts of the different ensemble members, the chaotic nature envisaged by Lorenz would be realised as of the growing differences in the forecast as related to differences in the initial conditions. This method would provide a means of assessing the sensitivity of weather prediction to errors or uncertainties in the initial conditions.

Secondly, different model predictions can be obtained using different model configurations. It is known that model formulations can be different in terms of dynamics, numerical method of solution and variations in horizontal and vertical resolution. For example, variations in model formulations such as grid-point or spectral method; different vertical coordinates such as sigma or Eta or Hybrid; and different resolutions.

Thirdly, differences in the representation of physical processes also contribute to uncertainty. There are now different parameterization schemes for each of the physical processes of radiation, convection, cloud microphysics, surface physics, etc due to differences in hypothesis and approximations involved in the development of a parameterization scheme. With the availability of multiple schemes for each of the physical process, multiple combinations of these would reflect as variation in the model formulation.

Fourthly, different boundary conditions would also lead to different predictions. The differences in the surface boundary values could arise from differences in the sea surface temperatures, soil moisture, soil type and vegetation; differences in defining the boundary conditions at the top of the model atmosphere such as rigid-top and gravity-drag approximation; and differences in the lateral boundary conditions (from ensemble predictions) in the case of limited area models.

As such, different model formulations and different parameterizations would generate different predictions starting from the same initial conditions. The outcome of this type of ensemble predictions with differences in model formulation and parameterizations would reflect the uncertainties associated with model imperfection. Since it may be nearly impossible to design and execute a perfect model that is capable of resolving all the atmospheric processes, it becomes imperative to understand the uncertainties associated with model imperfection.

Thus, the above described ensemble prediction method could be used to generate an ensemble of forecasts, and to ascertain the probabilities of individual forecasts and finally to propose the most probable forecast.

Herein, the strategies of generating ensembles from using perturbations in the initial conditions are presented. In the ensemble prediction system (EPS), the single deterministic forecast is determined starting from the best possible estimate of the initial conditions and this is referred to as "control" prediction. The ensemble members are then created by adding perturbations to the final analysis. It is imperative that the perturbations are compatible with the observed errors. A simple method of creating the ensembles is by adding random noise within the error limits, but this method has an inherent deficiency that the spatial errors may be correlated. For generating appropriate perturbations, an analysis of the growth of perturbations was considered. It is understood that there are essentially two types of errors (i.e.) arising due to fast-moving, low-amplitude, non-amplifying (random) and rapidly slow-moving, amplifying (organized) components in the analysis. Theoretically, even though both of these two modes start with nearly same amplitude, the non-amplifying modes (such as gravity waves) will get damped in contrast to the rapidly growing modes (baroclinically unstable) that will finally determine the error of a numerical forecast. Effectually, the domination of the non-amplifying errors will make the ensemble members to be closer to the "control" and if rapidly amplifying errors were to dominate then the "control" forecast will rapidly differ from the real atmosphere from the early stage of prediction. Thus, understanding and estimation of the rapidly-amplifying component and proper use of them would enable the majority of the ensemble member predictions to be nearer to the real atmosphere. The growth of errors will be fast as nearly exponential initially (at the beginning of model integration) but the error growth will reduce after about 2 weeks.

Considering the different strategies, the use of a pure random perturbation will lead to the growth of amplifying mode and decaying gravity-wave-modes but does not truly reflect the real uncertainties in the initial analysis. For ensuring better representation of the uncertainty, the single-vector and breeding methods were proposed.

Single Vector method: The single-vector method will be designed to have optimised perturbation growth over a defined time scale of a few days and this method has a limitation that it may sometimes give rise to unrealistic atmospheric structures. The European Center for Medium-range Weather Forecasts (ECMWF) uses the single-vector method, in which a simplified version of their model is used to generate 48-hour predictions, from which the error vectors are computed as the directions in which differences between control and forecast are increasing with time. Then the information of fast growing errors will be used to run the EPS backward to relate the initial condition uncertainty to the fast growing error vectors. These perturbation vectors will then be modified in magnitude considering the a priori information of the observation and first guess errors of the model's assimilation system. Lastly, these perturbations are added to the initial conditions of the forecast model.

Breeding method: National Centers for Environmental Prediction (NCEP) uses the breeding method for their EPS. The breeding method will permit the perturbations to be realistic and representative of the observed atmospheric structures. This method uses forward model integrations to generate and modify the perturbations. In this method perturbations are bred through a number of model integrations. Stepwise: (i) Random perturbations are added to first guess analysis and the control and perturbation forecast are produced. (ii) The differences between the control and perturbation forecast will be used to generate 3-dimensional perturbation. (iii) The magnitude of this perturbation is modified to reflect the uncertainties in the observations and model first guess. (iv) The new perturbations are then added to the initial guess of a new forecast from the next forecast time. Steps (i) to (iv) denote a breeding cycle, and the cycles are repeated till such a time that the differences between the control and forecast differences stabilises. After completion of the breeding cycles, the optimised perturbations that reflect the fast amplifying modes will be added to the control initial conditions to generate the members of the EPS. In this method, it is presumed that the fast growing modes will manifest and become dominant at the initial stages and retain their presence till the end of the forecast period. It is to be noted that the number of breeding cycles are dependent on the computational resources.

7. Current Status of NWP in India

Numerical Weather Prediction, in terms of model development, solution and application of output, had profoundly advanced due to advancement in computer technology (such as high performance computing), RADAR and Satellite observations, development of data assimilation methodologies and parallel processing algorithms. These advancements have transformed weather predictions on different time and spatial scales. As of the past, notable contributions have been made mainly from sources in USA and Europe followed by Japan, India and Australia. Due to the availability of resources, USA, European Union, Japan and India are generating weather predictions at short, medium, and long range time scales and at spatial scales of 3-km to 25-km. At the current time, different weather products from each of the above mentioned countries are being made available to the international community through web based public domains. Although many of these products are similar covering the entire globe, their model sources and technological applications differ.

In this concluding chapter of this monograph, an attempt is made to briefly present the salient features of one of these models and the weather products for information of reader. Where possible, the sources are given from which the reader will be able to extract more information as per their need.

United States of America: The National Centers for Environmental Prediction (NCEP) uses GFS global modeling system for generating global weather predictions.

The Global Forecast System (GFS)-Global Spectral Model (GSM) (GSM Version 13.0.2)

The National Centers for Environmental Prediction (NCEP) Global Forecast System (GFS) provides all operational numerical weather predictions, both deterministic and probabilistic, up to 16 days. The output from the global model provides necessary initial and boundary conditions for the regional, oceanic and wave prediction models. The atmospheric forecast model is a global spectral model run with T1534 (~13 km) and T574 (~34 km) resolutions in horizontal to generate predictions for time periods of 0-10 days and 10-16 days respectively. The vertical coordinate system uses 64 hybrid sigma- pressure levels with the model top at 0.27 hPa.

The operational dynamical core of the GFS is based on a two time-level semi-implicit semi-Lagrangian discretization with three dimensional Hermite interpolation (the dynamical core still supports three time-level Eulerian approach). The semi-Lagrangian advection calculations as well as treatment of physics are done on a linear, reduced (for computational economy) Gaussian grid in the horizontal domain. The semi-implicit treatment and implicit eighth order horizontal diffusion are performed in spectral space. To reduce noise in the stratosphere, additional second order divergence damping that increases with altitude is applied above ~100hPa. A correction is also applied to the global mean ozone to account for the non-conservation during semi-Lagrangian advection step.

Same physical parameterizations are used across all horizontal and vertical resolutions. The currently used longwave (LW) and the shortwave (SW) radiation parameterizations are the modified and optimized versions of the Rapid Radiative Transfer Models (RRTMG_LW v2.3 and RRTMG_SW v2.3, respectively).

For the planetary boundary layer (PBL) parameterization, eddy-diffusivity mass-flux (EDMF) parameterization is applied for the strongly unstable PBL, while the old GFS eddy-diffusivity counter-gradient parameterization is used for the weakly unstable PBL. For the vertical momentum mixing, the mass-flux parameterization that includes the effect of the updraft-induced pressure gradient force is used. The heating by turbulent kinetic energy (TKE) dissipation is parameterized to reduce an energy imbalance. To enhance a too weak vertical turbulent mixing for the weakly and moderately stable conditions, eddy-diffusivity profile method is used.

The gravity wave drag and mountain blocking parameterizations are currently applied as four times stronger mountain blocking and one half the strength of gravity wave drag than of the T382L64 version.

For shallow convection parameterization, a mass flux scheme based on the bulk mass-flux parameterization of deep convection is used. Separation of deep and shallow convection is determined by cloud depth (currently 150 hPa). Entrainment rate is given to be inversely proportional to height and much larger than that in the deep convection parameterization. Mass flux at cloud base is given as a function of the surface buoyancy flux.

For deep convection parameterization, the Simplified Arakawa-Schubert (SAS) that is revised to make cumulus convection stronger and deeper to reduce excessive grid-scale precipitation is used. Random cloud-top selection is replaced by an entrainment rate approach with an environmental moisture dependent entrainment rate. Convective overshooting and the effects of convection-induced pressure gradient force (which reduces convective momentum transport) are included. The cloud condensate is detrained from upper cloud layers above downdraft initiating level.

Currently, for land surface parameterization, the Noah LSM with four layers is used. This includes addition of frozen soil physics, new formulations for infiltration and runoff (giving more runoff for unsaturated soils), revised physics of the snowpack and its influence on surface heat flux and albedo, tuning and addition of canopy resistance parameters, spatially varying root depth, revised treatment of ground heat flux and soil thermal conductivity, reformulation for dependence of direct surface evaporation on first layer soil moisture, and improved seasonality of green vegetation cover. The initial conditions for the global forecasts are obtained through the Global Data Assimilation System (GDAS). The GDAS ingests all available global satellite, conventional (rawinsonde, aircraft, surface) and radar observations with a plus or minus 3:00 hour window of the analysis time. A 9-hour GSM forecast (T1534 interpolated to T574) from the previous GDAS analysis is used as the first guess for the assimilation. The GDAS runs with a late (6:00) data cut-off to provide the next 6 hourly cycle background using the largest amount of available observations. The GDAS uses a hybrid four-dimensional ensemble variational formulation, in which the propagation of the background error covariances in time is approximated by an hourly ensemble of forecasts rather than by a tangent linear and adjoint model as in 4DVar formulations. Within a variational framework, the hourly ensemble covariances are combined with a time-invariant estimate of the background error derived from the model's 24-48 hour forecast climatology.

Complete details of the NCEP GFS system is available at http://www.emc.ncep.noaa.gov/GFS/doc.php

NCEP is currently generating weather prediction data at 0.25^{0} horizontal resolution and at 47 vertical levels at four cycles per day starting at 00, 06, 12, 18 UTC with 15-days prediction and with predicted fields at 3 hour interval. In addition, NCEP also generate 21-member ensembles at 4 cycles per day.
The GFS global predicitons data are available at
http://www.ftp.ncep.noaa.gov/data/nccf/com/gfs/prod

GEFS (Global ensemble prediction system) ensemble data are available at
http://nomads.ncep.noaa.gov/pub/data/nccf/com/gens/prod/

As of the global predictions, limited area predicitons are equally important as the limited area models can be run at very high resolutions in order to predict mesoscale atmospheric phenomena which could not be predicted with coarser global models. WRF (Weather Research and Forecasting) modeling system is one of the limited area models, which is most widely used for research as well as operational purposes. Although there are several other models (such as RAMS, ARPS and UKMO unified model), WRF is the most popular with the widest use because of good documentation, portability on to different computer platforms and user-support. So the technical details of the WRF model are presented here in brief.

Weather Research and Forecasting Model

Weather Research and Forecasting (WRF) mesoscale atmospheric modeling system has been developed and sourced from National Center for Atmospheric Research (NCAR), as the next generation model after MM5, incorporating the advances in atmospheric simulation system suitable for a broad range of applications. WRF has two dynamical cores: The Advanced

Research WRF (ARW) and Nonhydrostatic Mesoscale Model (NMM). The dynamical core includes advection, pressure-gradients, Coriolis force, buoyancy, filters, diffusion, and time-stepping. Both are Eulerian mass dynamical cores with terrain-following vertical coordinates. ARW support and development are centered at NCAR/MMM whereas NMM development is centered at NCEP/EMC and support is provided by NCAR/DTC. Presently there are two WRF dynamical cores in the WRF framework, the NMM and ARW cores. Given their existing formulations, these two cores cannot be merged –the two cores differ fundamentally in a few crucial aspects.

An atmospheric dynamical core can be concisely defined by a relatively small number of attributes: form of the governing equations, prognostic variables, horizontal grid, vertical grid, terrain formulation, time integration method, spatial integration method.

The NMM and ARW cores possess the following attributes:

1. Governing equations formulation

NMM: Advective form for the momentum and thermodynamic equations with a mass continuity equation.

ARW: flux form for the momentum and thermodynamic equations with a mass continuity equation.

2. Prognostic variables

NMM: Velocities, temperature, pressure, and column mass (hydrostatic surface pressure).

ARW: Momentum (mass coupled velocities), dry entropy (mass coupled potential temperature), geopotential, and column mass (hydrostatic surface pressure).

3. Horizontal grid

NMM: E grid

ARW: C grid

4. Vertical grid

NMM and ARW – Lorenz grid

5. Terrain formulation

NMM: Terrain-following sigma coordinate that relaxes to a pressure coordinate at an intermediate level below the model top.

ARW: Terrain-following sigma coordinate.

6. Time integration method

NMM: Eulerian formulation. Forward-backward scheme for inertia-gravity waves, 2nd order Adams-Bashforth for horizontal advection and Coriolis terms, Crank-Nicholson scheme for vertical advection. Acoustic modes are integrated explicitly for horizontal propagation (using the model time step), and implicitly for vertical propagation.

ARW: Eulerian formulation. Forward-backward for inertia-gravity waves and horizontally propagating acoustic modes (time split, using an explicit acoustic small time step), 3rd order Runge-Kutta scheme for all other terms. Vertically propagating acoustic modes are integrated with an implicit scheme.

7. Spatial discretization

NMM: The horizontal discretization follows the enstrophy and energy conservation principles of Arakawa. The vertical discretization is centered and second order. Mass and moist scalar conservation is maintained in the discrete formulation. Horizontal divergence damping is used. The full thermodynamic variables are used for pressure-gradient calculations.

ARW: The horizontal discretization of the advection terms uses a 5th order upwind biased scheme. The mass-divergence and pressure gradient formulations are centered and 2nd order accurate. Vertical advection typically uses a third order upwind biased scheme. The discretization is in conservative form, with conservation of the first-order quantities – momentum, dry entropy, mass and moisture. Horizontal divergence damping is not used. The perturbation thermodynamic variables are used for pressure-gradient calculations.

The history of the two cores are as follows:

The NMM core: The nonhydrostatic NMM model evolved directly from the hydrostatic Eta model. Outside of the inclusion of the nonhydrostatic terms into the governing equations, the primary differences between the two (are the use of a terrain-following coordinate in the NMM versus the step-mountain coordinate in Eta, and the removal of the time-split gravity wave integration from the NMM that is used in the Eta model. Additionally, the NMM uses a 2nd order Adams-Bashforth time integration scheme as opposed to the first order scheme used in the Eta model. The configuration of the NMM has also evolved from that used for the Eta model. Specifically, the dissipation formulation and physics in the NMM have been modified and re-tuned to enhance and better retain the small scale structures in the forecasts. These small-scale structures are almost entirely missing from Eta forecasts and forecasts from early configurations of the NMM. The primary features shared between the NMM and Eta models are the use of the E grid and the use of the energy and enstrophy conserving finite differencing.

The ARW core: The ARW core has its roots in the Klemp-Wilhelmson (KW) cloud model developed in the mid and late seventies. The ARW uses the KW time-splitting of the acoustic and gravity waves. It differs from the KW model in most other aspects; the ARW core is cast in conservative form and the discrete model conserves mass, momentum, dry entropy and scalars; it uses a 3rd order Runge-Kutta time integration scheme for the slow-mode integration; it uses a hydrostatic pressure (mass) vertical coordinate; it uses high-order upwind biased flux-divergence (advection) operators. The KW formulation has been used as the basis for a number of other models, including MM5, COAMPS, ARPS, and the DWD-LM. These model formulations are very close in design to the original KW model. For example, all use an advective (non-conservative) formulation, and all use the KW leapfrog time-split integration method.

Similarities of the two cores: (i) First and foremost is that both the NMM and ARW cores share the hydrostatic pressure (mass) vertical coordinate – the NMM core retained it from its Eta model ancestry while the ARW model developed a new time split integration scheme for it; (ii) Both models conserve mass exactly - to machine round off; and (iii) Both use terrain-following coordinates. The difference in the formulations (the NMM relaxes to constant pressure at an intermediate level whereas ARW uses the sigma coordinate throughout the atmosphere) is not significant because the NMM formulation is easily included in the ARW core (a similar formulation is presently being considered). (iv) Both models use the forward-backward scheme for inertia gravity waves and horizontally propagating acoustic modes, and both use a vertically implicit formulation for vertically propagating acoustic modes. The details of the schemes differ because of the different variables and horizontal grid structures, but the philosophies behind their implementation and use are the same. (v) Both use the Lorenz grid for vertical discretization.

The differences between the two cores are: (i) Grid staggering, the NMM uses the E grid and the ARW uses the C grid. (ii) Choice of equations, variables and conservation properties: The NMM uses pressure and temperature so as to facilitate an energy and enstrophy conserving discretization, while the ARW core uses potential temperature to facilitate conservation of first order quantities (momentum, dry entropy). (iii) Time-integration methods: The NMM is a fully explicit model (except for the treatment of vertically propagating sound waves and vertical advection). The ARW core is a split-explicit model – acoustic and gravity waves are integrated explicitly with a small time step (using the forward-backward technique) while slow modes are integrated with a large time step (using 3rd order Runge-Kutta).

Herein, the particular properties of the ARW system only are provided as this core is more popularly used by the scientific community. ARW model system has versatility to choose the domain region of interest; horizontal resolution; interactive nested domains and with various options to choose parameterization schemes for physical processes. The model is designed to be a flexible, state-of-the-art atmospheric simulation system that is portable and efficient on available parallel computing platforms and a detailed description was provided by Skamarock et al. (2008).

The predicted variables are the velocity components u and v in Cartesian coordinate, vertical velocity w, perturbation potential temperature, perturbation geopotential, and perturbation surface pressure of dry air. Optionally, turbulent kinetic energy and any number of scalars such as water vapor mixing ratio, rain/snow mixing ratio, cloud water/ice mixing ratio, and chemical species and tracers.

The model has subgrid scale turbulence formulation in both coordinate and physical space that include divergence damping, external-mode filtering, and vertically implicit acoustic step off-centering. Initial conditions can be provided from real data or idealized data. Digital filtering initialization option is available for real data. Top level boundary conditions have options of rigid lid, gravity wave absorbing. The bottom boundary conditions have the physical or free-slip options. Four map projections are supported for real-data simulations as polar stereographic, Lambert conformal, Mercator and latitude-longitude. Curvature terms are included. Nesting options are available as one-way interactive and two-way interactive, and moving nests. The system has options for use of analysis and observation nudging. Option is there to make global simulation using polar filter and periodic east-west boundary conditions. The modeling system has a number of options for the parameterization schemes of the physical processes. The parameterizations of atmospheric radiation, planetary boundary layer, cloud microphysics, cumulus convection, and land surface process are included. The current version of WRF has WRF-Var merged into the framework, which facilitates data assimilation and ensemble formulation.

More details of WRF model documentation are available at http://www2.mmm.ucar.edu/wrf/users/docs/arw_v3.pdf and at http://www2.mmm.ucar.edu/wrf/users/docs/user_guide_V3.9/ARWUsersGuideV3.9.pdf

Current scenario of NWP in India

The Ministry of Earth Science (MoES) has broadly defined the goals and responsibilities of the 3 premier institutes of meteorology in India, namely the India Meteorological Department (IMD), the National Center for Medium Range Weather Forecasting (NCMRWF), and the Indian Institute of Tropical Meteorology (IITM). The current scientific activities of these institutes are briefly presented.

National Center for Medium Range Weather Forecasting: NCMRWF was started in 1988 with the main mandate of providing medium range weather forecasts (3 to 10 days in advance) and agro-meteorology advisories, using sophisticated numerical weather modelling techniques and powerful computing infrastructure, for the farming community in India. Accordingly, the first Super Computer System in India (Cray XMP/14, with 1 processor and 4 MB memory) was acquisitioned and dedicated to the Nation by the then Prime Minister of India in 1989. This ushered in the era of high-performance computing for Indian Meteorology.

In 1989, NCMRWF implemented the ECMWF global atmospheric model (Cycle-30 version), with an in-house developed global Optimum Interpolation (OI) data analysis scheme. Subsequently, COLA R40 model along with all components of NWP system viz., an OI based analysis scheme, diagnostic and graphics packages was implemented through Indo-US collaboration. During 1992-93, NCMRWF implemented global data assimilation forecast system at T80L18 resolution of NCEP, USA along with upgrading of data analysis and assimilation procedure from OI to Spectral Statistical Interpolation (SSI) scheme. Statistical Dynamical models based on global NWP outputs were developed to generate the location specific forecasts for different agro-climatic zones of India. The global forecast model T80L18 (~150 km horizontal resolution and 18 vertical layers) was made operational in June 1994. After successful development and demonstration of AAS for the agro-climatic zones, NCMRWF transferred the responsibility to IMD in 2007.

NCMRWF was involved in the research, performed several sensitivity experiments related to the physical parameterization schemes of PBL, Radiation, Convection and cloud parameterisation and orography to improve the dynamic core of the modelling system. Subsequently a Regional Spectral Model (RSM) based on Perturbation Technique developed at NCEP was nested with the operational global T80L18 model and implemented. NCMRWF also implemented Eta model (a step-mountain eta (coordinate model) of NCEP in 2001, which was first run on-real time basis for issuing weather forecasts for the Indian Army's Everest Expedition during April-May of 2001. The Eta model (48km/38L resolution) along with the NCMRWF operational T80L18 global model analysis was made operational (for 72 hrs forecast) since January 2002. The MM5 model (NCAR, USA) was also installed and tested with double and triple nested domains at 90, 30 and 10 km resolutions over the Indian region. The innermost domain at 10 km resolution was kept over the Jammu & Kashmir region to support the Government of India initiated PARWAT programme. MM5 model was subsequently replaced by WRF models (WRF-ARW & WRF-NMM) in 2005.

Subsequently, in 2002 the weather forecast model was upgraded to T170L28 (~75km). A new global high-resolution assimilation-forecast system GFS (Global Forecast System) at T254L64 (~50km) resolution was implemented operationally in 2007 along with assimilation of radiance observations from various global polar-orbiting satellites. Subsequently, the SSI data assimilation system was replaced by Grid-point Statistical Interpolation (GSI) scheme and the GFS was further upgraded to T574L64 (~22 km) resolution in 2010 and T1534L64 (~12km) resolution in 2016.

The GSI scheme of data assimilation was upgraded to Hybrid GSI with ETR ensembles (based on T190/T254) in 2015 and with EnKF ensembles (based on T574) in 2016. Subsequently the horizontal resolution of Hybrid-Ensemble (with 80 members) system has been increased to 12 km (T1534L64). Scientists at NCMRWF have developed and implemented the assimilation of various new Indian satellite observations viz., Ocean surface winds from Oceansat-2 Scatterometer (OSCAT), Radiances from MT-SAPHIR microwave radiometer and ROSA radio-occultation, INSAT-3D atmospheric motion vectors and radiances, ScatSat-1, etc. The real time initial conditions generated through this system are being used by IMD and IITM for running their operational global deterministic as well as ensemble prediction system. A 12-year (2000-2011) GFS based reanalysis was carried out at T574L64 resolution at NCMRWF. This reanalysis product is widely used for seasonal and extended range predictions by IMD, IITM and INCOIS.

NCMRWF started developing a system for seamless prediction of weather and climate wherein the same modelling system with advanced data assimilation techniques is used across a range of spatial and temporal scales. Under a special MoU with UK Met Office, Unified Model (UM) at 25 km horizontal resolution along with 4D-Variational Data Assimilation scheme was first implemented in 2012. NCMRWF, in-house developed modules to pack the observations in to a form compatible to UM system. Indian satellite observations are also being assimilated in this system. Subsequently this seamless prediction system was upgraded in 2015 to a 17-km global model, a 4-km regional model and a very high resolution 1.5 km model (specifically for resolving and capturing severe weather events).

In order to predict visibility during winter over Delhi, a 330-m regional model which uses the 30-m CartoSat DEM was configured and successfully trialled during the winter season of 2016. An Extended Kalman Filter based land data assimilation system was developed for assimilation of soil moisture. Six hourly soil moisture analysis, at four soil levels covering 3 metre depth, produced by this system are used to initialize the soil moisture conditions. This land data assimilation system is capable of assimilating the soil and skin temperatures as well. An algorithm for quality characterization in terms of Quality Index Maps for the radial wind velocity and the radar reflectivity of Indian Doppler Weather Radar (DWR) has also been developed. NCMRWF initiated the efforts to assimilate Indian DWR products in to the regional data assimilation system.

A 20 member NCMRWF Global Ensemble Forecast System at 70 km horizontal resolution, based on GFS, was implemented and made operational in 2012 for generating probabilistic forecasts for the first time in the country. Probabilistic Quantitative Precipitation Forecast (PQPF) for various rainfall thresholds and the strike probability of tropical cyclones were made available for the first time in India. This system was further upgraded by implementing the bias correction techniques in 2014. Subsequently, this version of the ensemble prediction system was transferred to IITM.

A high resolution (33 km) global ensemble prediction system based on UM model was developed and implemented in October 2015. This 44 member NCMRWF Global Ensemble Prediction System uses NCMRWF's UM (NCUM) operational deterministic model analysis as control initial conditions. The perturbations are generated using Ensemble Transform Kalman Filter method. Based on the 10-day ensemble member forecasts, NCUM started generating several products like ensemble mean, spread, PQPF, EPSgrams, tropical cyclone ensemble tracks, strike probability, probability of snow etc.

Acquiring and sustaining supercomputing infrastructure was the most critical part of the NCMRWF objectives. After first successful acquisition of Cray-XMP/14, the same system was upgraded in March 1992 to a dual processor system Cray XMP/216. To utilize indigenously developed technology in India, distributed memory systems were installed at NCMWRF in 1999 and all the parallelised versions of the weather codes were ported on these systems. Subsequently NCMRWF procured two shared memory systems, a 24-processor Cray SV1 system in 2001 and 64 processors Cray X1E in 2006. In the process of upgrading its computing resources, 24 Tera Flops and 350 Tera Flops IBM systems were installed at NCMRWF in 2010 and 2015 respectively. Work is already underway to enhance this computing power further to 2.8 PetaFlop along with 6 PB of storage in 2018.

NCMRWF has started developing a very high resolution (~12 km) NCUM based global ensemble prediction system for generating very high resolution probabilistic forecasts up to 10 days. This system will be configured based on several sensitivity experiments based on ensemble member size, model resolution, etc. Evaluation of these systems will be carried out by hindcast of extreme weather events during recent past. Various bias corrected forecast products based on this system will be developed. NCMRWF is developing high resolution coupled modelling system along with coupled data assimilation scheme. This system will be helpful for accurate prediction of monsoon features with a lead time of two weeks. NCMRWF is also developing various customized NWP applications for energy sectors.

Ocean surface parameters (wind and fluxes) from global and mesoscale models of NCMRWF were used for driving many of the wave and ocean models of the country in both real-time and research mode which played a pivotal role in the country's Ocean State Forecasting (OSF) efforts.

Data assimilation was one of the major important activities at NCMRWF since its inception. The scientists of the centre conducted several Observations System Experiments (OSE) and validation studies using conventional and satellite observations, especially observations from Indian satellite viz. INSAT, Kalpana, and IRS-P4/OCEANSAT-1. These OSEs were very crucial for improving the data assimilation over Indian region. NCMRWF also developed and implemented a real-time observation monitoring system with special emphasis on Indian satellite observation.

NCMRWF currently uses the new 350 tera flop (TF) IBM iDataPlex HPC system named "Bhaskara" for its research and development. At the current time, NCMRWF is entrusted with the responsibility of generating suitable grid point values covering the entire globe to be used by IMD and IITM for their prediction applications. NCMRWF uses the 4D-Var data assimilation methodology to generate the global initial conditions.

The 4D-Var system uses the background fields and linearization state for assimilating the observations from different sources. Essentially, global observations from different sources and the background error fields are assimilated into the model forecast based guess fields to generate the analysis increments.

India Meteorological Department

IMD, with the active support of IITM and NCMRWF operates state-of-the-art NWP and climate models, starting from mesoscale model for Nowcasting to short rage weather forecast (up to 3 days), global model for medium range weather forecast (up to 10 days), Global Ensemble Prediction System for probabilistic forecasts and Global Coupled GCM (CGCM) for extended range forecast (up to a month time scale) and seasonal forecast. The operational products are available at http://nwp.imd.gov.in/

Short Range Weather Forecasting: IMD uses Doppler Weather Radar (DWR) observations for nowcastng of local weather and also for their assimilation into high resolution numerical modeling system for mesoscale forecasting. IMD is using the Warning Decision Support System Integrated Information (WDSSII), as developed by National Severe Storm Lab, USA, for nowcasting since 2010. Around the same time, IMD also started the use of Advanced Regional Prediction System (ARPS) at 9 km resolution with the assimilation of DWR data. IMD started the use of nonhydrostatic version of MM5 (Mesoscale Model Generation 5) model for operational predictions in the year 2005, and the new generation non-hydrostatic WRF (Weather Research and Forecast) modelling system at 27-km resolution since 2006, which was made fully operational in 2009. IMD improved the WRF modeling system with double nested configuration of 27 and 9 km domains, and implementation of 3-D variational data assimilation to generate forecasts for 3 days. The WRF model along with data assimilation component (WRFDA), with improved physics at 9 km resolution is made operational since 2015, utilising the High Performance Computing System "Aaditya" at IITM, Pune. Following the nest-down approach, real-time predictions at 3-km resolution were started in 2016 on experimental basis, and made operational in 2017 generating 1-hour interval forecasts up to 3-days.

Medium Range Weather Forecasting: IMD started using NCEP based Global Forecast System (GFS T382) in 2009 to produce operational predictions for 7-days at two cycles per day (00 UTC and 12 UTC). IMD has upgraded the model system to GFS T574; (~25 km horizontal resolution) in 2012 and to GFS T1534 (~12 km horizontal resolution) from 2016 with model forecasts up to 10 days. Bias corrected technique was implemented for surface maximum and minimum temperature and the outputs are used to generate meteograms at 5-days time scale for 650 cities. In addition, IMD had started generating 20-member ensemble predictions in 2017, using the global ensemble forecast system (GEFS) at T574 resolution for medium range weather forecasting. These modelling systems were implemented at IMD in collaboration with IITM and NCMRWF under the national monsoon mission.

Observational aspects for NWP: For the purpose of generating improved initial conditions for application to the short and medium range forecasts (forecast up to 7 days in tropics), IMD has taken up the process of expansion and digitization of its observational network to achieve seamless data flow consolidating data from Doppler Weather Radars (DWR), Satellites (INSAT-3D Radiance), wind profilers, GPS sonde, meso-network (Automatic Weather Stations), buoys and aircrafts in the real time mode which are ingested into the assimilation cycle of global and mesoscale NWP models.

Agro-meteorology advisories: Considering the need of farming sector, IMD has upgraded the Agro-Meteorological Advisory Service from agro climate zone to district level, and started issuing district level weather forecasts up to 5-days from 1 June 2008. These forecasts are generated through Multi-Model Ensemble (MME) system making use of model outputs from state-of-the-art global models from the leading global NWP centres. This has considerably improved the accuracy and reliability of NWP product although limitations remained regarding the prediction of intensity and mesoscale rainfall distribution that cause inland flooding.

Extended Range Weather Forecasting: IMD started providing extended-range forecasts using the real time coupled model outputs available from the National Centre for Environmental Prediction (NCEP),USA; the European Centre for Medium Range Weather Forecasting (ECMWF); and the Japan Meteorological Agency (JMA), Japan. Since July, 2016, IMD started generating operational Extended Range Forecast products using CFSv2 coupled model, which was tested and developed at IITM and transferred to IMD. IMD also started using a suite of models at different resolutions, which are (i) CFSv2 at T382 (≈ 38 km) (ii) CFSv2 at T126 (≈ 100 km) (iii) GFSbc (bias corrected SST from CFSv2) at T382 and (iv) GFSbc at T126 with the atmospheric and oceanic Initial conditions obtained from NCMRWF and INCOIS assimilation systems respectively. The operational suite of models is now being used for day-to-day operational prediction on the ADITYA HPCS at IITM. The Multi-model ensemble (MME) of the above 4 suite of models are run operationally for 32 days based on every Wednesday, initial condition with 4 ensemble members (one control and 3 perturbed) each for CFSv2T382, CFSv2T126, GFSbcT382 and GFSbcT126, to generate the mean and anomaly forecast valid for 4 weeks for days 3-9 (week1; Friday to Thursday), days 10-16 (week2; Friday to Thursday), days 17-23 (week3; Friday to Thursday) and days 24-30 (week4; Friday to Thursday).

Long Range Forecasting: IMD has established an experimental dynamical forecasting system, based on Experimental Climate Prediction Center (ECPC) seasonal forecasting model (SFM), for the long range forecasting of Indian summer monsoon rainfall at IMD-Pune centre. The SFM model (AGCM) is being used to prepare monthly and seasonal forecasts during the monsoon season using persistent and forecasted SSTs as boundary forcing. The experimental forecasting system has shown useful forecast skill. Currently, efforts are being made to implement the fully coupled model (CFSv2) for the seasonal forecast under the Monsoon Mission.

IMD, along with numerical model based predictions, is using statistical techniques to provide operational long range forecasts. The statistical approach has shown good skill in generating forecasts for the seasonal rainfall for the Indian subcontinent as a whole and for the four geographical regions (Northwest India, Central India, Northeast India and South Peninsula) and monthly rainfall for the Indian subcontinent for the months of July, August and September. However, IMD has reported limited skill for monthly and seasonal forecasts for smaller spatial scales such as state, subdivision, district etc.

Prediction of tropical cyclones: A cyclone specific forecasting model HWRF (Hurricane WRF) was introduced into the NWP framework of IMD in 2011 in collaboration with NOAA as a part of INDO-US TC-Project through an MOU between MoES and NOAA. The model, with its double nested (27 km × 9 km) configuration, had been made operational in 2012. The HWRF modelling system has been upgraded further with improved physics and vortex initialization in 2014 and triple nested configuration (27 km × 9 km × 3 km), and with a moving option for the inner nests. The model was further improved in 2016, with the increase of the resolution of the model (18 km × 6 km × 2 km) and with sophisticated hybrid-EnKF GSI assimilation.

IMD has also started preparing the cyclone Genesis Potential Parameter (GPP) and multi-model ensemble based track forecast in the medium range time scale using the output data from NWP models. IMD had also identified that a threshold of GPP ≥ 8.0 indicate its potential for intensification (T. No. 1.0 to 2.0) into a cyclone at early stages of development. A dynamical statistical model (SLIP) is also developed in-house for Intensity prediction in five days time scale.

Weather Prediction over polar region: Polar-WRF model has been implemented for generating weather predictions over the Maitri region in Antarctica up to 48 hours since 2011. The model uses a single static domain with (400 x 400) grids at 15 km horizontal resolution using polar-stereographic projection and 39 vertical Eta levels with the location of Maitri station (70o46' S, 11o44' E) at the center of the model domain, and with the initial and boundary fields taken from IMD's operational global GFS-574 model. Recently, the model has been ungraded with improved model physics and increased horizontal resolution of 9 km to provide hourly forecasts for both the Maitri and Bharati (69o24' S, 76o11' E) stations of India in Antarctica region.

Indian Institute of Tropical Meteorology

IITM has been the backbone of research in India for meteorology, continuously providing support to NWP at the India Meteorological Department. Researchers at IITM have extensively contributed to the understanding of various meteorological phenomena including the monsoons, tropical cyclones, atmospheric modelling etc.

Research on northward propagation of summer monsoon low-frequency intraseasonal modes simulation by global atmospheric models was initiated by Krishnan and Kasture in the later 90s. Investigations of global climate and monsoons at the Climate and Global Modeling Division (CGMD) using global Atmospheric General Circulation Models (AGCM) was a significant steps in the modelling activities of IITM. Several experiments were carried out using the AGCMs from UK Met

Office and Center for Ocean-Land-Atmosphere Studies (COLA) towards understanding the Indian summer monsoon dynamics and variability. The studies have identified that the combined influence of the suppressed monsoon convection and the Himalayan orographic barrier played an important role in anchoring the quasi-stationary mid-latitude troughs over west-central Asia during monsoon-drought periods. Sensitivity to the parameterization of moist convective processes was made with both the AGCMs and limited area models. Research at IITM with GCMs have produced many important results in the understanding of Indian monsoon systems and its predictability.

Some of the notable contributions are: IITM model simulations were made a part of the AMIP (Atmospheric Model Intercomparison Project) international program for assessment of simulations of 1997-98 ENSO event and the mean Asian summer monsoon and its intra-seasonal variability. IITM has taken up the Seasonal Prediction of the Indian Monsoon (SPIM) project in which the skill of UKMO and COLA AGCMs were evaluated. These studies have brought out that the model biases in the simulated monsoon precipitation are due to sensitivity to SST anomalies associated with ENSO variations and inability to simulate the link with the Equatorial Indian Ocean Oscillation (EQUINOO)

Their research with GCMs have conclusively found that persistently warmer-than-normal SST in the equatorial Indian Ocean contributes to weakening of the Hadley circulation that leads to prolonged "breaks" in the Indian monsoon rainfall.

Long-period (43-year) simulations of a multi-level global ocean GCM have lead to improved understanding of ENSO-monsoon dynamical linkage, indicating that the external forcing of anomalous SST warming in the "central equatorial Pacific" is most effective in inducing monsoon-droughts over India.

GCM simulations at IITM have also showed that internal feedbacks between the monsoon convection and the mid-latitude circulation produce extended monsoon-breaks leading to severe drought-conditions over the Indian subcontinent.

Based on the strong modeling research initiatives, IITM took up mission mode modeling activities during 2012-2017, which lead to the development of an Earth System Model (ESM), which is the first climate model from India to contribute to the CMIP (Coupled Model Intercomparison Project) experiments and the IPCC (Intergovernmental Panel on Climate Change) assessment process. In the area of regional climate change research, an ensemble of high resolution downscaled projections of regional climate and monsoon until 2100 for the IPCC climate scenarios (RCP4.5 and RCP8.5) using a regional climate model (ICTP-RegCM4) at 50 km resolution have been evolved by dynamical downscaling. These results have also been made part of the Coordinated Regional Downscaling Experiment (CORDEX) South Asia activity [http://cccr.tropmet.res.in/cordex/files/data_on_esgf.jsp].

Under the umbrella of National Monsoon Mission, a sponsored program of Ministry of Earth Sciences (MoES), IITM has achieved a landmark in the monsoon forecast and research. Through this mission IITM has setup a state of the art dynamical modeling frame work for improving prediction skill of Seasonal and Extended range predictions and Short and Medium range (up to two weeks) predictions. Efforts are being made for seamless prediction of monsoon and climate under the National Monsoon Mission. Under this mission activity, the prediction skill of monsoon over India using a CFS (Climate Forecast System) model at (T382L64 (~38 km) horizontal resolution). The CFS based multi model ensemble prediction system, involving runs with ECMWF TL255 (~ 77 km) UKMO N216 (~ 50 km) and NCEP T126 (~110 km), is being used to predict the active/break cycles of monsoon with reliable prediction skill up to 3 pentads and skilful guidance up to 4 pentads over the country.

IITM has also setup a high resolution T574 (semi-lagrangian core) Global Ensemble Forecast System (GEFS) with 21 ensemble members, to provide real-time short range ensemble forecast since June 2016. This forecast system has been recently transferred to IMD for operational implementation.

A very high resolution T1534 (~12.5 km horizontal resolution) deterministic forecast system has also been set up since August

2016 for short range predictions using GFS (Global Forecast System model). Efforts are presently underway to setup ensemble prediction system based on GEFS at T1534 (~12.5 km) resolution as a commitment to Niti-Aayog. The deterministic GFS T 1534 model has also been transferred to IMD for operational implementation. Thus, IITM has accomplished a major target of developing and enhancing National forecasting system through improved modeling system. Future plans: The IITM ESM will be participating in the International Panel for Global Climate Change (IPCC) Coupled Model Inter-comparison Project (CMIP-6) during 2020. This will be the first ever effort from India to participate in the IPCC assessment. A plan has been made to develop a high-resolution version of IITM-ESM (atmosphere grid size: 38 km, ocean grid: 0.5 deg x 0.5 deg and 0.25 deg x 0.25 deg near equator) for long-term climate studies and generate finer resolution projections of future climate and monsoon until 2100 for different climate scenarios. IITM also plans to develop a next-generation IITM-ESM by incorporating more Earth System components, including interactive aerosols and trace gases and land biogeochemistry.

IITM plans to develop applications, based on the current modeling system, particularly short/seasonal/extended range forecasts to water, power, agriculture and hydrology etc. Setting up a fully coupled data assimilation system for improved atmosphere and ocean initial conditions for short and long range forecasts is another planned activity. Further improvement of the present Monsoon Mission Model (CFS) through developmental activities carried out by various national/international partners would be another activity to be accomplished in near future. IITM has plans to develop a seamless prediction system using regional coupled model, and incorporating new dynamical core in the present modeling system to evolve a very high resolution model to improve prediction of weather extremes with more accuracy.

IITM produces real time seasonal forecasts of the Indian summer monsoon and other phenomena such as El Nino and Indian Ocean Dipole. IITM uses the high resolution version of current climate system version 2 ($CFSV_2$) to generate the predictions at approximate horizontal resolution of 38 km. The forecasts are generated using an ensemble of 40 members, initialized once in a month. IITM also performs a part of the extended range predictions for use by IMD.

Indian Air Force

The Indian Air Force established the Centre for Numerical Weather Prediction (AFCNWP) in 1990 to meet their specific requirements of weather forecasts. The AFCNWP with its location in New Delhi functions directly under the Directorate of Meteorology of IAF in New Delhi, has been carrying out operational NWP activities for Indian Air Force (IAF). AFCNWP is a nerve centre, which looks after all the weather related requirements i.e. operational as well as research and development of Indian Air Force. The centre is functioning round the clock for enabling real time weather exchange of the entire Indian Air Force and providing guidance in long, medium and short range forecast over complete Indian subcontinent and all the regions of interest for planning and conduct of operations. The centre also provides meteorological and NWP guidance for VVIP / VIP Movements, International / National exercises, Adventure Activities and National Events including Relief and Rescue Operations. The other major tasks of AFCNWP is to maintain climatology of India and ex-India, train Air Force personnel and naval personnel on NWP and associated software and participate and support National / International Projects.

AFCNWP initiated their activity of data archival and developing met-softwares with the acquisition of a computer. As a part of it, the Centre for Operational Meteorological and Extended Territorial Data Integrated Software Knowledge (COMET DISK) was formed in New Delhi In 2002, with the aim to generate meteorology utility software and manage meteorological data. The first operational use of MM5 model in the IAF was demonstrated during the international event 'Aero-India 2005' in Feb 2005 at Air Force Station, Yehalanka (Bangalore). AFCNWP started running the MM5 model (at 81 km resolution) since 2005, and at 81 and 27 km resolution since 2006.

AFCNWP got equipped with its first High performance Computing System in October 2009, and started using the WRF ARW 3.1.1 model for operational predictions since December 2009. The WRF-ARW model is being run at 27-18-6 km resolution since 2012 and the model products are being provided to the field units as NWP guidance for their day to day forecasting.

Since its formation, the AFCNWP has provided forecasts for various exercises within the IAF as well as civil meteorological agencies. An experiment to predict convective events over Northwest India, named 'Megh Garjan 2006', was conducted by the centre during pre-monsoon season of 2006. The centre took active part in national projects 'STORM' and 'PRWONAM' to predict severe thunderstorms over Northeast India. It has actively participated in many of programs of IMD and ISRO. It has generated forecasts for various IAF movements to foreign countries inclusive of the Micro-light expedition in 2007 and exercise Red Flag, which displayed Indian Air power in various foreign countries in the year 2015.

In the year 2006, "Mausam-online" - a single point meteorological data entry and real-time visualization package was developed by AFCNWP. This package caters the entry of the Current Weather data, Local Forecasts, TAFs, Warnings and upper winds for IAF Met Sections. After initiation of the Mausam website and WRF model, NWP products were provided as a guidance for day to day forecasting. Mausam-Online version 2.0 was launched in August 2015, which received the appreciation of the Honorable Prime Minister of India, Mr. Narender Modi in October, 2015.

The AFCNWP has been recently equipped with a new HPCS (30 Teraflop and 200 Terabite), which enhanced the computational capabilities by manifold and the centre presently is carrying out operational NWP twice a day using the WRF ARW 3.7.1 model at horizontal resolutions of 9 and 3 km for the forecast periods of 7 and 3 days respectively. The centre generates guidance for various aviation weather hazards using latest techniques. This centre is manned by highly trained and efficient IAF personnel and functions round the clock.

The AFCNWP also houses various equipments, such as the Microwave Radiometer, the Lightning Detection system and the Doppler Weather Radar servers. The server for the "Mausam Online" website, which is the life line of all the meteorological sections is also placed here. The maintenance and updation of this site is done by the personnel of AFCNWP.

In order to stay ahead in the ever growing field of mesoscale prediction, the meteorological branch of the IAF is rapidly gaining expertise in fine tuning of various models and operational usage of products towards accurate weather predictions. The AFCNWP is acting as the hub centre in provisioning numerical predictions, especially in respect of the mesoscale weather phenomena and aims to become centre of excellence in the field of meso-scale weather forecasts with effective coordination with other organisations in this field (IMD, IIT, NCMRWF etc.).

Other institutions: Research in NWP is actively pursued at many of the Universities, IITs, IISc and national laboratories the results of which find application in the operational predictions from the India Meteorological Department.

Appendix – A: Mathematical Formulation of NWP

At first, it is necessary to ascertain the use and application of the fundamental governing equations of atmospheric motion for the purpose of weather prediction.

The equations of motion are

$$\frac{du}{dt} - fv + \frac{1}{\rho}\frac{\partial p}{\partial x} = 0 \qquad \text{.....(A.1)}$$

$$\frac{dv}{dt} - fv + \frac{1}{\rho}\frac{\partial p}{\partial x} = 0 \qquad \text{.....(A.2)}$$

$$\frac{\partial p}{\partial z} + g\,\rho = 0 \qquad \text{.....(A.3)}$$

The equation of continuity is

$$\frac{\partial \rho}{\partial t} + u\frac{\partial \rho}{\partial x} + v\frac{\partial \rho}{\partial y} + w\frac{\partial \rho}{\partial z} + \rho\left(\frac{\partial u}{\partial x} + \frac{\partial v}{\partial y} + \frac{\partial w}{\partial z}\right) = 0 \qquad \text{.....(A.4)}$$

The equation of state is

$$p\,\alpha = RT \text{ or } p = R\rho T \qquad \text{.....(A.5)}$$

The thermodynamic energy equation (First law of Thermodynamics for adiabatic process)

$$\frac{d}{dt}\theta = \frac{d}{dt}T\left(\frac{p_o}{p}\right)^k = 0 \qquad \text{.....(A.6)}$$

$$\theta = T\left(\frac{p_o}{p}\right)^k \qquad \text{.....(A.7)}$$

where k = R/Cp

There are seven equations and seven atmospheric variables, which are u, v, w, p, ρ, T and θ that denote a closed system. However the prediction algorithm is not direct, as the computations of 'w' and 'p' are not explicit. The prediction process can be arrived at with some rearrangement of the equations.

Equations (A.3), (A.4) and (A.7) can be combined as follows

$$\frac{\partial p}{\partial z} = -g\rho$$

$$P = \rho RT$$

where $k = R/C_p$

$$\theta = T\left(\frac{p_o}{p}\right)^k$$

$$\frac{\partial p}{\partial z} = -\frac{g\rho}{RT}$$

$$\frac{T}{P}\frac{\partial p}{\partial z} = -\frac{g}{R}$$

$$\frac{\theta}{P}\left(\frac{p}{P_o}\right)^k \frac{\partial p}{\partial z} = -g/R$$

$$\frac{\left(\frac{p}{p_o}\right)^k}{\rho}\frac{\partial p}{\partial z} = -g/R\theta$$

$$\frac{\partial}{\partial z}\left(\frac{p}{p_o}\right)^k = -\frac{gk}{R\theta}$$

Integrating with respect to z, we get

$$\left(\frac{p}{p_o}\right)^k = -\int\frac{gk}{R\theta}\,\partial z$$

$$= -\frac{gk}{R}\int\frac{1}{\theta(z)}\,dz \qquad \text{.....(A.8)}$$

This equation denotes the derivation of p (pressure) at any level (height) z using the information of the vertical variation of θ (potential temperature).

Next, the equation of continuity (equation A.4) can be modified as follows

$$\frac{\partial \rho}{\partial t} + u\frac{\partial \rho}{\partial x} + v\frac{\partial \rho}{\partial y} + w\frac{\partial \rho}{\partial z} + \rho\left(\frac{\partial u}{\partial x} + \frac{\partial v}{\partial y} + \frac{\partial w}{\partial z}\right) = 0$$

Substituting for

$$\rho = -\frac{1}{g}\frac{\partial p}{\partial z}$$

$$\frac{\partial \rho}{\partial t} = -\frac{1}{g}\frac{\partial}{\partial t}\left(\frac{\partial p}{\partial z}\right) = -\frac{1}{g}\frac{\partial}{\partial z}\left(\frac{\partial p}{\partial t}\right)$$

$$\frac{\partial \rho}{\partial x} = \frac{\partial}{\partial x}\left(-\frac{1}{g}\frac{\partial p}{\partial z}\right) = -\frac{1}{g}\frac{\partial}{\partial z}\left(\frac{\partial p}{\partial x}\right)$$

$$u\frac{\partial \rho}{\partial x} = \frac{u}{g}\frac{\partial}{\partial z}\left(\frac{\partial p}{\partial x}\right)$$

$$\frac{\partial}{\partial z} = \left(u\frac{\partial p}{\partial x}\right) = \frac{\partial u}{\partial z}\frac{\partial p}{\partial x} + u\frac{\partial}{\partial z}\frac{\partial p}{\partial x}$$

$$\therefore u\frac{\partial \rho}{\partial x} = -\frac{1}{g}\left(\frac{\partial}{\partial z}\left(u\frac{\partial p}{\partial x}\right)\right) + \frac{1}{g}\frac{\partial u}{\partial z}\frac{\partial p}{\partial x}$$

Similarly

$$v\frac{\partial \rho}{\partial y} = -\frac{v}{g}\frac{\partial}{\partial z}\left(\frac{\partial p}{\partial y}\right) = -\frac{v}{g}\frac{\partial}{\partial y}\left(\frac{\partial p}{\partial z}\right)$$

$$= -\frac{1}{g}\left(\frac{\partial}{\partial z}\left(v\frac{\partial p}{\partial y}\right)\right) = +\frac{1}{g}\frac{\partial v}{\partial z}\frac{\partial p}{\partial y}$$

$$w\frac{\partial \rho}{\partial y} = -\frac{1}{g}w\frac{\partial}{\partial z}\left(\frac{\partial p}{\partial z}\right)$$

$$= -\frac{1}{g}\left(\frac{\partial}{\partial z}\left(w\frac{\partial p}{\partial z}\right)\right) + \frac{1}{g}\frac{\partial w}{\partial z}\frac{\partial p}{\partial z}$$

$$\frac{\partial}{\partial z}\left(\frac{\partial p}{\partial t} + u\frac{\partial p}{\partial x} + v\frac{\partial p}{\partial y} + w\frac{\partial p}{\partial z}\right)$$

$$-\frac{\partial u}{\partial z}\frac{\partial p}{\partial x}-\frac{\partial v}{\partial z}\frac{\partial p}{\partial y}-\frac{\partial w}{\partial z}\frac{\partial p}{\partial z}+\frac{\partial p}{\partial z}\nabla.V=0$$

$$\frac{\partial}{\partial z}\left(\frac{dp}{dt}\right)-\frac{\partial v}{\partial z}.\nabla p+\frac{\partial p}{\partial z}\nabla.V=0$$

Integrating with respect to z

$$\frac{dp}{dt}=-\int_{z}^{\infty}\left(\frac{\partial V}{\partial z}.\nabla p+\frac{\partial p}{\partial z}\nabla.V\right)dz \qquad .....(A.9)$$

The 1st law of thermodynamics, which is written as

$$C_v\, dT + p\, d\alpha = 0$$

and the equation of state, which is expressed as

$$p\,\alpha = RT$$

are combined to give [using $\gamma = C_p/C_v$; and $R = C_p - C_v$]

$$\frac{1}{\gamma p}\frac{dp}{dt}=\frac{1}{\rho}\frac{dp}{dt} \qquad .....(A.10)$$

The equation of continuity is written as

$$\frac{1}{\rho}\frac{d\rho}{dt}+\nabla.V\frac{\partial w}{\partial z}=0$$

$$\frac{\partial w}{\partial z}=-\nabla.V-\frac{1}{\rho}\frac{d\rho}{dt}==-\nabla.V-\frac{1}{\gamma p}\frac{dp}{dt}$$

substituting equation (A.10) for $\frac{dp}{dt}$

$$\frac{\partial w}{\partial z}=-\nabla.V+\frac{1}{\gamma p}\int_{z}^{\infty}\left(\frac{\partial V}{\partial z}.\nabla p+\frac{\partial p}{\partial z}\nabla.V\right)dz$$

Integrating with respect to z

$$w = -\int_0^Z (\nabla.V)dz + \frac{1}{\gamma}\int_0^Z (\frac{1}{p})$$

$$\int_z^\infty (\frac{\partial V}{\partial z}.\nabla p - \frac{\partial P}{\partial Z}\nabla.V)\ dz'\ z'' \qquad .....(A.11)$$

This equation shows the derivation of vertical velocity 'w' from the vertical distributions of horizontal divergence and vertical gradients of velocity and pressure.

Now the final equations are written as

$$\frac{\partial u}{\partial t} = -u\frac{\partial u}{\partial x} - v\frac{\partial u}{\partial y} - w\frac{\partial u}{\partial z} - fv + \frac{1}{\rho}\frac{\partial p}{\partial x} \qquad .....(A.12)$$

$$\frac{dv}{dt} = fu + \frac{1}{p}\frac{\partial p}{\partial y} \qquad .....(A.13)$$

$$\frac{d\theta}{dt} = 0 \qquad .....(A.14)$$

$$\rho = \frac{p}{RT} \qquad .....(A.15)$$

$$\theta = T\left(\frac{p_o}{p}\right)^k \quad \text{or} \quad T = \theta\left(\frac{p}{p_o}\right)^k \qquad .....(A.16)$$

$$\left(\frac{p}{p_o}\right)^k = \frac{Kg}{R}\int_z^\infty \frac{1}{\theta(z)}dz$$

$$w = -\int_0^Z (\nabla.V)dz + \frac{1}{\gamma}$$

$$\int_0^Z (\frac{1}{p})\int_z^\infty (\frac{\partial V}{\partial z}\nabla.p - \frac{\partial P}{\partial Z}\nabla.V)dz^1\ dz^{11} \qquad .....(A.17)$$

Equations (A.12) - (A.18) provide a closed set of equations with the seven variables of u, v, w, p, ρ, T and θ.

Equations A.12, A.13 and A.14 are three prediction equations while the remaining four (A.15, A.16, A.17 and A.18) are diagnostic equations.

To assess the prediction system, observations of u, v, p and T only are available and are to be used at the start time. The following are the computational steps to be followed:

Step 1: Observations of p and T are used to compute the 3-D distribution of θ and ρ

Step 2: u, v, and p are used to compute the distribution of w (vertical velocity)

Step 3: This completes the production of the distributions of all the seven variables at the start time.

For integration forward, to derive the predicted variables, the following procedure is adopted.

Step 1: The three prediction equations A.12, A.13 and A.14 are used to obtain the rate of change of variables u, v and θ and so the future values (i.e.) after a chosen time step.

Step 2: The vertical distribution of θ is used to obtain the pressure values of p.

Step 3: The values of p and θ are used to obtain temperature T.

Step 4: p and T values are used to calculate ρ.

Step 5: Using u, v and p, the vertical velocity w is obtained. These steps are repeated to obtain the future values of all the seven variables u, v, w, p, ρ, T and θ.

The above description provides the basic mathematic formulation of weather prediction. Theoretically, it shows that starting at some initial time with the proper description of the initial state (3-dimensional observations of the four variables) of the atmosphere, these equations could be used to obtain the future state of the atmosphere by integrating the equations in forward mode.

Although, the problem of weather prediction could be expressed mathematically, a careful evaluation brings out some issues in their solution. First and foremost issue is how to solve the equations. These are nonlinear partial differential equations, which imply that certain terms are nonlinear. In particular, these are the advection terms which involve the product of two differently varying terms such as "u" and $\frac{\partial u}{\partial x}$. The nonlinear nature restricts direct analytical solution, in contrast to the direct solutions for the linear differential equations. This necessitates finding alternate methods, and "finite difference" methods are identified to be most suitable for their solution.

Appendix – B: Sigma Coordinate System

In meteorology, for the purpose of atmospheric model development, where variations in topography are important, a terrain-following vertical coordinate has been proposed by Philips (1957).

In this a variable "sigma σ" is defined as pressure normalized to its surface value. A simple definition is $\sigma = \frac{p}{p_s}$ where p_s is surface pressure

This denotes σ =1 at the surface and σ =0 at the top of atmosphere where p =0

Alternatively $\sigma = \frac{p - p_T}{p_s - p_T}$

Where p_T is the pressure at the top of model atmosphere.

It can be easily seen that $\sigma' \,(\sigma - vertical\ velocity) = \frac{d\sigma}{dt}$

σ' = 0 at σ =1 (surface)

Now the transformation from p-coordinates to σ - coordinate system will be elaborated.

The equation of motion is

$$\frac{dV}{dt} + fk * V = -\nabla\phi$$

where

$$\frac{d}{dt} = \frac{d}{dt} + V.\nabla + \omega \frac{d}{dp}$$

Where in the horizontal operator corresponds to differential on an isobaric surface.

Since σ and p surfaces do not coincide, the transformation can be made as similar to transformation from z- to p-coordinates

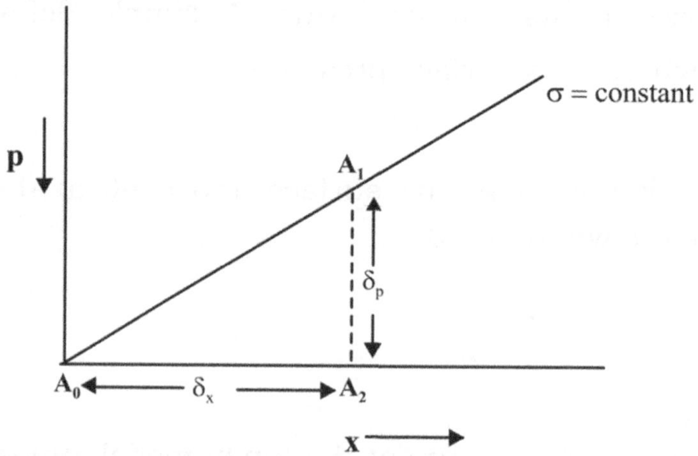

Figure A.1 Representation of relation between p- and σ- surfaces

where 'A' denotes any variable

$$\frac{A_1 - A_o}{\delta x} = \frac{A_2 - A_o}{\delta x} - \frac{A_2 - A_1}{\delta x}$$

$$= \frac{A_2 - A_o}{\delta x} = \frac{A_2 - A_1}{\delta \sigma} - \frac{\delta \sigma}{\delta x}$$

where $\delta\sigma = \left(\frac{\delta\sigma}{\delta x}\right)_p \delta x$

$$\left(\frac{\partial A}{\partial x}\right)_\sigma = \left(\frac{\partial A}{\partial x}\right)_p - \frac{\partial A}{\partial \sigma}\left(\frac{\partial \sigma}{\partial x}\right)_p$$

(p_s is variant in x direction)

$$\left(\frac{\partial\sigma}{\partial x}\right) = \frac{\partial(p/p_s)}{\partial x} = -\frac{p}{p_s^2}\frac{\partial p_s}{\partial x} = -\frac{\sigma}{p_s}\frac{\partial p_s}{\partial x}$$

$$\left(\frac{\partial A}{\partial x}\right)_\sigma = \left(\frac{\partial A}{\partial x}\right)_p + \frac{\sigma}{p_s}\frac{\partial p_s}{\partial x}\frac{\partial A}{\partial\sigma}$$

$$\left(\frac{\partial A}{\partial x}\right)_p = \left(\frac{\partial A}{\partial x}\right)_\sigma - \frac{\sigma}{p_s}\frac{\partial p_s}{\partial x}\frac{\partial A}{\partial\sigma}$$

Similarly we can write

$$\left(\frac{\partial A}{\partial y}\right)_p = \left(\frac{\partial A}{\partial y}\right)_\sigma - \frac{\sigma}{p_s}\frac{\partial p_s}{\partial y}\frac{\partial A}{\partial\sigma}$$

In vector form

$$\nabla_p(\ ) = \nabla_\sigma(\ ) - \frac{\sigma}{p_s}\nabla p_s\frac{\partial(\ )}{\partial\sigma}$$

The equation of motion , in σ – coordinates is

$$\frac{dV}{dt} + fk \times V = -\nabla\phi + \frac{\sigma}{p_s}\nabla p_s \nabla\phi$$

where ϕ is geopotential and $\frac{d}{dt} = \frac{\partial}{\partial t} + V.\nabla + \sigma'\frac{\partial}{\partial\sigma}$

The equation of continuity in p- coordinates

$$\frac{\partial u}{\partial x} + \frac{\partial v}{\partial y} + \frac{\partial w}{\partial p} = 0$$

is now written as

$$\frac{\partial}{\partial p} = \frac{\partial}{\partial p_s \sigma} = \frac{1}{p_s}\frac{\partial}{\partial\sigma}$$

$$(\nabla.\mathrm{V})_p + \frac{1}{p_s}\frac{\partial\omega}{\partial\sigma} = 0$$

The σ' (σ - Vertical Velocity) is defined as

$$\sigma' = \frac{d\sigma}{dt} = \left[\frac{\partial\sigma}{\partial t} + V.\nabla_\sigma\right] + \omega\frac{\partial\sigma}{\partial p}$$

$$\frac{\partial\sigma}{\partial t} = \frac{\partial}{\partial t}(\frac{p}{p_s}) = -\frac{\mathrm{p}}{\mathrm{p_s}^2}\frac{\partial p_s}{\partial t} = -\frac{\sigma}{\mathrm{p_s}}\frac{\partial p_s}{\partial t}$$

$$\frac{\partial\sigma}{\partial t} = \frac{\partial}{\partial t}(\frac{p}{p_s}) = -\frac{\mathrm{p}}{\mathrm{p_s^2}}\frac{\partial p_s}{\partial t} = -\frac{\sigma}{\mathrm{p_s}}\frac{\partial p_s}{\partial t}$$

Similarly

$$\mathrm{V}.\nabla\sigma = -\frac{\sigma}{p_s}\mathrm{V}.\nabla p_s$$

$$\therefore \sigma' = -\frac{\sigma}{p_s}\left[\frac{\partial p_s}{\partial t} + V.\nabla p_s\right] + \frac{\omega}{p_s}$$

$$(\nabla.V)_p = (\nabla.V)_\sigma - \frac{\sigma}{p_s}\nabla p_s\frac{\partial V}{\partial\sigma}$$

$$\frac{\omega}{p_s} = \sigma'\frac{\sigma}{p_s}\left[\frac{\partial p_s}{\partial t} + V.\nabla p_s\right]$$

$$\omega = \sigma' p_s + \sigma\left[\frac{\partial p_s}{\partial t} + V.\nabla p_s\right]$$

$$\frac{\partial\omega}{\partial\sigma} = \frac{\partial}{\partial\sigma}\sigma' p_s + \frac{\partial}{\partial\sigma}\sigma\left[\frac{\partial p_s}{\partial t} + V.\nabla p_s\right]$$

$$\nabla.(p_s V) + p_s\frac{\partial\sigma'}{\partial\sigma} + \frac{\partial p_s}{\partial t} = 0$$

$$\frac{\partial\omega}{\partial\sigma} = \frac{\partial(\sigma' p_s)}{\partial\sigma} + \frac{\partial}{\partial\sigma}\sigma\left(\frac{\partial p}{\partial t} + V.\nabla p_s\right)$$

$$\left[\frac{\partial p_s}{\partial t} + V.\nabla p_s\right]_p = \left[\frac{\partial p_s}{\partial t} + V.\nabla p_s\right]_\sigma$$

$$\dot{\sigma} = \frac{\partial \sigma}{dt} = \left[\frac{\partial \sigma}{\partial t} + V.\nabla\sigma\right]_{\sigma} + \dot{\sigma}\frac{\partial \sigma}{\partial \sigma}$$

$$\frac{\partial \dot{\sigma} p_s}{\partial t} = \dot{\sigma}\frac{\partial p_s}{\partial \sigma} + p_s\frac{\partial \dot{\sigma}}{\partial \sigma} = p_s\frac{\partial \dot{\sigma}}{\partial \sigma}$$

[2nd term on RHS is zero because P_s does not change with σ]

$$\frac{\partial}{\partial \sigma}\sigma\frac{\partial p_s}{\partial t} = \frac{\partial \sigma}{\partial \sigma}\frac{\partial p_s}{\partial t} + \sigma\frac{\partial}{\partial \sigma}\frac{\partial p_s}{\partial t}$$

$$= \frac{\partial p_s}{\partial t} \textit{ on a sigma surface}$$

$$\frac{\partial}{\partial \sigma}\sigma V.\nabla p_s = \frac{\partial \sigma}{\partial \sigma}V.\nabla p_s + \sigma\frac{\partial}{\partial \sigma}V.\nabla p_s$$

$$= V.\nabla p_s + \sigma\frac{\partial}{\partial \sigma}V.\nabla p_s$$

$$\therefore \frac{\partial \omega}{\partial \sigma} = p_s\frac{\partial \dot{\sigma}}{\partial \sigma} + (\frac{\partial p_s}{\partial t})_{\sigma} + (V.\nabla p_s)_{\sigma} + \sigma\frac{\partial V}{\partial \sigma}.\nabla P_s$$

Substituting this expression and

$$(\nabla.V)_p = (\nabla.V)_{\sigma} - \frac{\sigma}{p_s}\frac{\partial V}{\partial \sigma}.\nabla p_s$$

Into

$$p_s.(\nabla.V)_p + \frac{\partial \omega}{\partial \sigma} = 0 \qquad \textit{(equation for continuity)}$$

We will have

$$p_s.(\nabla.V)_{\sigma} + p_s\frac{\partial \dot{\sigma}}{\partial \sigma} + \frac{\partial p_s}{\partial t} + V.\nabla p_s = 0$$

$$\frac{\partial p_s}{\partial t} + p_s\frac{\partial \dot{\sigma}}{\partial \sigma} + p_s\nabla V. + V.\nabla p_s = 0$$

$$\frac{\partial p_s}{\partial t} + p_s\frac{\partial \dot{\sigma}}{\partial \sigma} + \nabla.p_s V = 0$$

Integrating this equation with respect to σ with the boundary conditions of σ =0 at p=P_s and p=P_T, we get

$$\frac{\partial p_s}{\partial t} = -\int_0^1 \nabla . p_s \mathrm{V}\, \partial\sigma$$

The equation of state: P=R ρT

Hydrostatic equation $\frac{\partial p}{\partial z} = -gp = -\frac{gp}{RT}$

$$\mathrm{g}\,\partial z = -\frac{RT}{p}dp$$

$$\partial\phi = \frac{RT}{p}dp = -\frac{RT}{p}p_s\partial\sigma$$

$$\frac{\partial\phi}{\partial\sigma} = -\frac{RT}{\sigma} \quad (\because \sigma = \frac{p}{p_s})$$

Similarly

$$\theta = \mathrm{T}\left(\frac{p_o}{p}\right)^{\mathrm{R/C_p}} = \mathrm{T}\left(\frac{p_o}{\sigma p_s}\right)^{\mathrm{R/C_p}}$$

The equations in σ coordinates are now written as follows
The equation of motion

$$\frac{dV}{dt} + fk \times V = -\nabla\phi + \frac{\sigma}{p_s}\nabla p_s \nabla\phi \qquad \text{.....(B.1)}$$

Equation of continuity

$$\nabla.(p_s V) + p_s\frac{\partial\sigma'}{\partial\sigma} + \frac{\partial p_s}{\partial t} = 0 \qquad \text{.....(B.2)}$$

Hydrostatic equation

$$\frac{\partial\phi}{\partial\sigma} = -\frac{\mathrm{RT}}{\sigma} \qquad \text{.....(B.3)}$$

Potential temperature

$$\theta = T\left(\frac{p_o}{\sigma p_s}\right)^{R/C_p} \qquad .....(B.4)$$

Thermodynamic energy equation

$$\frac{D\theta}{Dt} = \frac{\partial\theta}{\partial t} + (V.\theta)_\sigma + \sigma'\frac{\partial\theta}{\partial\sigma} = Q \qquad .....(B.5)$$

Surface pressure tendency equation

$$\frac{\partial p_s}{\partial t} = -\int_0^1 \nabla.(p_s V)d\sigma \qquad .....(B.6)$$

The above set of six equations with the six variables v , σ , θ , T , ϕ and P_s constitute the closed system of primitive equations

Appendix – C: Finite Difference Formulation of Multi-Level Primitive Equation Model

The model equations are written as

$$\frac{\partial u}{\partial t}+\sigma'\frac{\partial u}{\partial \sigma}+m\left(u\frac{\partial u}{\partial x}+v\frac{\partial v}{\partial y}+\frac{\partial\, gz}{\partial x}+c_p\theta\frac{\partial \pi}{\partial x}\right)$$

$$-v\left(f-v\frac{\partial m}{\partial x}+u\frac{\partial m}{\partial y}\right)+Fx=0 \qquad .....(C.1)$$

$$\frac{\partial v}{\partial t}+\sigma'\frac{\partial v}{\partial \sigma}+m\left(u\frac{\partial u}{\partial x}+v\frac{\partial v}{\partial y}+\frac{\partial\, gz}{\partial y}+c_p\theta\frac{\partial \pi}{\partial y}\right)$$

$$-u\left(f-v\frac{\partial m}{\partial x}+u\frac{\partial m}{\partial y}\right)+Fx=0 \qquad .....(C.2)$$

$$\frac{\partial\, gz}{\partial \sigma}+C_p\theta\frac{\partial \pi}{\partial \sigma}=0 \qquad .....(C.3)$$

$$\frac{\partial \theta}{\partial t}+\sigma'\frac{\partial \theta}{\partial \sigma}+m\left(n\frac{\partial \theta}{\partial x}+v\frac{\partial \theta}{\partial y}\right)+H=0 \qquad .....(C.4)$$

$$\frac{\partial}{\partial t}\left(\frac{\partial p}{\partial \sigma}\right)+\frac{\partial}{\partial \sigma}\left(\sigma'\frac{\partial p}{\partial \sigma}\right)+m\left[\frac{\partial}{\partial x}\left(u\frac{\partial p}{\partial \sigma}\right)+\frac{\partial}{\partial y}\left(v\frac{\partial p}{\partial \sigma}\right)\right]$$

$$-\frac{\partial p}{\partial \sigma}\left(u\frac{\partial m}{\partial x}+v\frac{\partial m}{\partial y}\right)=0 \qquad .....(C.5)$$

The finite difference formulation is represented as

$$f_x=\frac{\partial f}{\partial x}=\frac{f_{i+1/2}-f_{i-1/2}}{x_{i+1/2}-x_{i-1/2}} \qquad ....\ \text{differencing}$$

$$f^{-x}=\frac{f_{i+1/2}+f_{i-1/2}}{2} \qquad ....\ \text{averaging}$$

$$f_{xx}=\frac{\partial}{\partial x}\left(\frac{\partial f}{\partial x}\right)=\frac{\dfrac{\partial f_{i+1/2}}{\partial x}-\dfrac{\partial f_{i-1/2}}{\partial x}}{x_{i+1/2}-x_{i-1/2}}$$

$$= \left[\frac{f_{i+1} - f_i}{x_{i+1} - x_i} - \frac{f_i - f_{i-1}}{x_i - x_{i-1}}\right]\frac{1}{x_{i+\frac{1}{2}} - x_{i-\frac{1}{2}}}$$

$$= [\frac{f_{i+1} - 2f_i - f_{i-1}}{\left(x_{i+\frac{1}{2}} - x_{i-\frac{1}{2}}\right)}$$

$$f^{-xx} = \left[\frac{f_{i+1} + f_i}{2} + \frac{f_i + f_{i-1}}{2}\right]^{\frac{1}{2}}$$

$$= \frac{f_{i+1} + 2f_i + f_{i-1}}{4}$$

$$f_x^{-x} = \left[\frac{f_{i+1} + f_i}{2} - \frac{f_i + f_{i-1}}{2}\right]\frac{1}{x_{i+\frac{1}{2}} - x_{i-\frac{1}{2}}}$$

$$= \frac{f_{i+1} - f_{i-1}}{2\left(x_{i+\frac{1}{2}} - x_{i-\frac{1}{2}}\right)} = \frac{f_{i+1} - f_{i-1}}{x_{i+1} - x_{i-1}}$$

In this formulation, staggering of variables is adopted within the 8 points of a cube

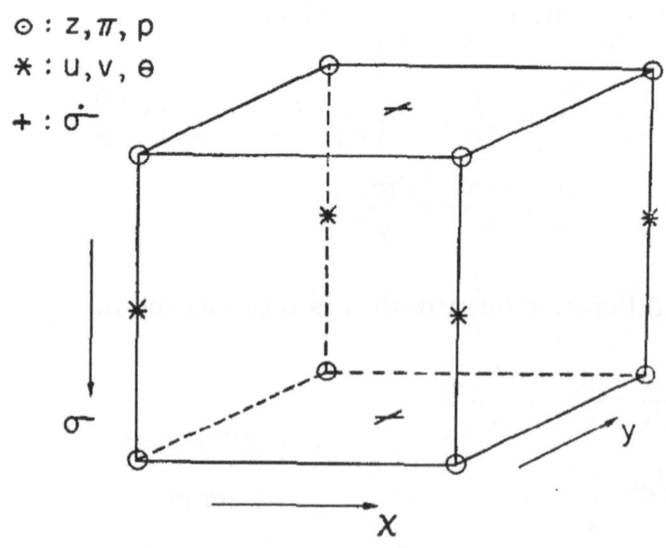

C.1 Schematic diagram showing the connections of the eight grid points in 3-dimensions. z, π and p are computed at the corner points; u, v and Θ are computed at midpoints of the four

vertical lines; and σ' is computed at the centre of the upper and lower faces.

z, π, p at circled points

u, v, Θ at star points

$\dot{\sigma}$ at plus points

In order to apply and solve the model equations (C.1) to (C.2), averaging of variables is needed. So the scheme of averaging and differencing used by Shuman and Hover mole (1968) are as follows.

Adopting the notation for differencing and averaging the five governing equations are written as

$$u_t^{-t} + [\sigma' u_\sigma^{-t} - V^{-xy}(f^{-xy} - v^{-xy} + u^{-xy})F_x]$$

$$+[u^{-xy}u_x^{-y} + v^{-xy}u_y^{-x} + g\bar{z}_x^{-\sigma y} + Cp\theta^{-xy}\bar{\pi}_x^{-\sigma y} = 0 \qquad \text{.....(C.6)}$$

$$v_t^{-t} + [\sigma' v_\sigma^{-t} + u^{-xy}(f^{-xy} - v^{-xy} + u^{-xy}]F_y]$$

$$+[u^{-xy}v_x^{-y} + v^{-xy}v_y^{-x} + g\bar{z}_x^{-\sigma y} + Cp\theta^{-xy}\bar{\pi}_y^{-\sigma x}] = 0 \qquad \text{.....(C.7)}$$

$$gz_\sigma + Cp\theta\pi_\sigma = 0 \qquad \text{.....(C.8)}$$

$$\theta_t^{-t} + [\sigma'\bar{\theta}_\sigma^{-xy} + u^{-xy}\bar{\theta}_\sigma^{-xy} + v^{-xy}\bar{\theta}_\sigma^{-xy} + H] = 0 \qquad \text{.....(C.9)}$$

$$P_{\sigma t}^{-t} + [\bar{P}_\sigma^{-xy}\sigma'_\sigma + u^{-xy}\bar{P}_{\sigma x}^{-y} + v^{-xy}\bar{P}_{\sigma y}^{-y}]$$

$$+[\bar{P}_\sigma^{-xy}(\bar{u}_x^{-xy} + \bar{v}_y^{-xy}) - u^{-xy} - v^{-xy}] = 0 \qquad \text{.....(C.10)}$$

Please note that the term "m" (map factor) is intentionally omitted to present the equations in a simpler form.

References

Baer, F., 1977: "Adjustment of initial conditions required to suppress gravity oscillations in nonlinear flows". Beit. Phys. Atmos., 50, 350–366.

Barnes, S. L., 1964: "A technique for maximizing details in numerical weather-map analysis". Journal of Applied Meteorology. 3 (4): 396–409.

Bourke, W., 1974: "A multi-level spectral model. I. Formulation and hemispheric integrations". Mon. Wea. Rev., 102, 687–701.

Bushby, F. H. and Timpson, M. S., 1967: "A 10-level atmospheric model and frontal rain". Q. J. Roy. Meteor. Soc., 93, 1–17.

Charney, J. G., 1947: "The dynamics of long waves in a baroclinic westerly current". J. Meteor., 4, 135–162.

Charney, J. G., 1948: "On the Scale of Atmospheric Motions". Geofysisk. Publ., 17, No. 2, 17 pp.

Charney, J. G., 1949: "On a physical basis for numerical prediction of large-scale motions in the atmosphere". J. Meteor., 6, 371–385.

Charney, J. G., Fjørtoft, R., and von Neumann, J., 1950: "Numerical integration of the barotropic vorticity equation". Tellus, 2, 237–254.

Charney, J. G., 1955: "The use of the primitive equations of motion in numerical prediction". Tellus, 7, 22–26.

Cressman, G.P., 1959: "An operational objective analysis system". Monthly Weather Review, Vol. 87, pp. 367–374.

Daley, Roger, 1991: "Atmospheric Data Analysis". Cambridge Univ. Press. 457 pp.

Dickinson, R., and Williamson, D., 1972: "Free oscillations of a discrete stratified fluid with application to numerical weather prediction". J. Atmos. Sci., 29, 623–640.

Eliassen, A., 1949: "The quasi-static equations of motion with pressure as independent variable". Geofysisk. Publ., 17, No. 3, 44 pp.

Eliasen, E., Machenhauer B., and Rasmussen, E., 1970: "On a numerical method for integration of the hydrodynamical equations with a spectral representation of the horizontal fields". Report No. 2, Institute for Theoretical Meteorology, University of Copenhagen.

Exner, F. M., 1908: U ber eine erste Anna¨herung zur Vorausberechnung synoptischerWetterkarten. Meteor. Zeit., 25, 57–67. Translated from German as: A first approach towards calculating synoptic forecast charts, with a biographical note on Exner by Lisa Shields and an Introduction by Peter Lynch. Historical Note No. 1, Met ´ Eireann, Dublin, 1995, 31 pp.

Fjørtoft, Ragnar, 1952: "On a numerical method of integrating the barotropic vorticity equa-tion". Tellus, 4, 179–194.

Hinkelmann, K., 1959: "Ein numerisches Experiment mit den primitiven Gleichungen". In The Atmosphere and the Sea in Motion. Rossby Memorial Volume, Rockerfeller lnsti-tute Press, 486–500.

Krishnamurti, T.N., Xue, J., Bedi, H.S., Ingles, K., Oosterhof, D., 1991: "Physical initialization for numerical weather prediction over the tropics". Tellus, 43AB, 53–81.

Lorenz, E. N., 1963: "Deterministic non-periodic flow". J. Atmos. Sci. 20, 130.

Lynch, Peter and Xiang-Yu Huang, 1992: "Initialization of the HIRLAM model using a digital filter". Mon. Weather Rev., 120, 1019–1034.

Machenhauer, B., and R. Daley, 1972: "A baroclinic primitive equation model with a spectral representation in three dimensions". Report No. 4, Institute of Theoretical Meteorology, University of Copenhagen, 63 pp.

Machenhauer, B., 1977: "On the dynamics of gravity oscillations in a shallow water model" with application to nonlinear normal mode initialization.' Beitrage zur Physik der Atmosphare, 50, 253–271.

Margules, M., 1904: "Uber die Beziehung zwischen Barometerschwankungen und Konti-nuit¨atsgleichung". Ludwig Boltzmann Festschrift. Leipzig, J A Barth. 930 pp. [On the relationship between barometric variations and the continuity equation. Translation in Lynch, 2001.]

Mesinger F. and A.Arakawa. (1976). Numerical Models Used in Atmospheric Models, 170p. GARP Publication Series, No 17, WMO/ICSU, Vol 1.

Miyakoda, K. and Moyer, R., 1968: "A method of initialization for dynamical weather fore-casting". Tellus, 20, 115–128.

Nitta, T. and Hovermale, J., 1969: "A technique of objective analysis and initialization for the primitive forecast equations". Mon. Weather Rev., 97, 652–658.

Orszag, S. A., 1970: "Transform method for calculation of vector-coupled sums: Application to the spectral form of the vorticity equation". J. Atmos. Sci., 27, 890–895.

Panofsky, H.A., 1949: "Objective weather map analysis". J. Met., Vol. 6, No.6, pp. 386–392.

Persson, A., 2005: Early operational numerical weather prediction outside the USA: an historical introduction. Part 1: Internationalism and engineering NWP in Sweden, 1952–69. Meteor. Appl., 12, 135–160.

Phillips, N. A., 1960: On the problem of initial data for the primitive equations. Tellus, Vol. 12, Issue 2, p.121

Phillips, N. A., 1990: The emergence of quasi-geostrophic theory. Pp. 177–206 in Lindzen et al. (1990).

Platzman, G. W., 1967: A retrospective view of Richardson's book on weather prediction. Bull. Amer. Met. Soc., 48, 514–550.

Reiser, H., 2000: The development of numerical weather prediction in the Deutscher Wetterdienst. In 50th Anniversary of Numerical Weather Prediction. Commemorative Symposium. Arne Spekat, Editor. Deutsche Meteorologische Gesellschaft, pp. 51–78.

Richardson, Lewis F., 1922: Weather Prediction by Numerical Process. Cambridge Uni-versity Press, xii+236 pp. Reprinted by Dover Publications, New York, 1965, with a new Introduction by Sydney Chapman, xvi+236 pp.

Rossby, C.-G. and collaborators, 1939: Relations between variations in the intensity of the zonal circulation of the atmosphere and the displacements of the semipermanent centers of action. J. Marine Res., 2, 38–55.

Sasaki, Y., 1958: An objective analysis based on the variational method. J. Met. Soc. Japan, 36, 77–88.

Shuman, F. G., 1989: History of numerical weather prediction at the National Meteorolog-ical Center. Wea. & Forecast., 4, 286–296.

Silberman, I., 1954: 'Planetary waves in the atmosphere. J. Meteor., 11, 27–34.

Slingo, J. and T. Palmer., 2011: Uncertainty in weather and climate prediction. Phil. Trans. R. Soc. A, 369, 4751–4767.

Smagorinsky, J., 1958: On the numerical integration of the primitive equations of motion for baroclinic flow in a closed region. Mon. Weather Rev., 86, 457–466.

For Product Safety Concerns and Information please contact our EU representative GPSR@taylorandfrancis.com Taylor & Francis Verlag GmbH, Kaufingerstraße 24, 80331 München, Germany